LEADING HIGH-STAKES MODERNIZATION IN THE AGE OF AI

START AT ZERO

ABJIHIT VEREKAR

Start at Zero: Leading High-Stakes Modernization in the Age of AI

2026 by Abhijit Verekar

ISBN:
Paperback: 979-8-9954535-0-5
eBook: 979-8-9954535-1-2
Audiobook: 979-8-9954535-2-9

First Edition

Contents

Foreword

Technology decisions in the public sector carry a weight that few private organizations experience. When a government system fails, the consequences extend far beyond balance sheets or quarterly targets. Payroll may not run for thousands of public employees. Utilities may struggle to bill residents. Permits may stop moving. Essential services that communities depend on every day can slow, stall, or collapse.

For that reason, modernization in government is never just a technology project. It is a matter of public trust.

Across cities, counties, states, and public agencies, leaders are now confronting one of the largest waves of technology change in decades. Legacy enterprise systems that have supported government operations for twenty years or more are reaching the end of their lifecycle. Vendors are pushing organizations toward cloud platforms. Artificial intelligence is beginning to reshape how information is analyzed, how systems are configured, and how decisions are made.

These forces are creating a moment of both opportunity and risk.

On one hand, governments now have access to tools that can dramatically improve how they operate. Modern platforms can connect departments that once worked in isolation. Data can move more quickly between systems. AI can help organizations understand requirements, detect risks, and manage complex implementations in ways that were difficult only a few years ago.

On the other hand, these same forces are driving thousands of public organizations into high stakes modernization decisions at the same time. Many of these entities have not undertaken a project of this scale in fifteen or even twenty years. The leaders responsible for these decisions often inherit legacy systems they did not choose, constraints they did not create, and timelines they did not set.

In this environment, the pressure to move quickly can be intense.

Vendor presentations promise transformation. Implementation roadmaps suggest predictable progress. Yet the history of large public sector technology programs tells a different story. Delays, cost over-runs, and operational disruption have become familiar headlines. In many cases, the organizations involved were led by capable and dedicated public servants who simply did not have the structural advantages needed to succeed.

The uncomfortable reality is that many failures begin long before implementation.

By the time contracts are signed and project teams are mobilized, the conditions that shape success or failure are often already in place. Requirements may have been shaped by vendor assumptions. Governance structures may not provide real independence. Oversight mechanisms may be too weak to challenge decisions once momentum builds.

When those conditions exist, even the most capable teams can struggle.

This is where the perspective behind this book matters.

Abhijit has spent years working alongside the people responsible for implementing these systems. He has sat in the same conference rooms as city managers, CIOs, and finance leaders wrestling with decisions that will shape their organizations for decades. He has seen projects from the inside, not only when they are celebrated at kickoff but when they face the difficult realities of implementation.

That vantage point is rare.

Many books about technology modernization are written either by vendors promoting their platforms or by observers analyzing outcomes from a distance. Abhijit writes from a different position. He has walked beside the teams responsible for delivering these transformations. He has seen the pressure they face, the information gaps they must navigate, and the structural forces that quietly shape the decisions they make.

Just as importantly, he has seen many different implementations unfold.

Some succeeded because leaders invested early in clarity, governance, and independence. Others struggled because preparation was rushed and key decisions were shaped by forces the organization did not fully understand. Across these experiences, patterns emerged. Success was rarely about a single product or methodology. It was about structure, incentives, and preparation.

This book captures those lessons.

At its core is a deceptively simple idea. If organizations want better outcomes from technology modernization, they must begin earlier than most do. They must create clarity before procurement begins.

They must define their needs before vendors define them. They must build governance structures before the pressures of implementation make independent oversight difficult.

The author calls this Phase Zero.

Phase Zero is not a technical milestone. It is the disciplined work of preparation that happens before the first vendor presentation and before the first contract is signed. It is the process of understanding the organization itself, its processes, its priorities, and its constraints before technology decisions lock those realities into place.

In government, this preparation is not optional. It is essential.

Public institutions operate under constraints that private companies rarely face. They must balance transparency with efficiency, accountability with speed, and long term stewardship with immediate service delivery. Decisions must withstand public scrutiny, legislative oversight, and leadership transitions. At the same time, the systems being modernized often support functions that cannot pause while transformation takes place.

In this environment, preparation becomes a form of risk management for the public good.

What makes this book particularly valuable is that it translates experience into practical guidance. The insights presented here come from observing and participating in real modernization efforts where decisions produced real consequences for organizations and the communities they serve.

Across those experiences, one pattern becomes clear.

Successful modernization projects do not begin with software. They begin with structure.

They begin with leaders who take the time to understand the systems they are responsible for changing.

They begin with organizations that define their needs before vendors define them.

They begin with governance models that ensure advice is independent and decisions are transparent.

Most importantly, they begin with the discipline to slow down at the beginning so that progress can accelerate later.

That discipline can be difficult in public institutions where urgency is constant. Communities expect better services, faster systems, and more responsive government. Leaders feel pressure to deliver results quickly.

Yet modernization done poorly can cause more disruption than the legacy systems it replaces. Rushing into procurement without clarity often leads to years of course corrections, contract renegotiations, and operational instability.

The message of this book is not that governments should avoid modernization. Quite the opposite. Technology transformation is essential for institutions that want to serve citizens effectively in a digital world.

The message is that modernization must begin with preparation.

For government executives, this book offers a practical guide to navigating one of the most consequential decisions they will face in their careers. For public sector technology professionals, it provides frameworks and language that help explain structural risks that are often difficult to articulate within large institutions. For elected officials and oversight bodies, it offers insight into the forces that shape outcomes long before problems become visible to the public.

Most importantly, it reinforces a principle that applies to every successful modernization effort.

Technology alone does not transform institutions. Sound decisions do.

Those decisions require clarity, independence, and preparation.

They require leaders who recognize that the most important work often happens before the project officially begins.

They require the willingness to begin in the one place many organizations overlook.

At zero.

Vidhu Shekhar

Chief Data and Artificial Intelligence Officer, Caltrans

Why This Book, Why Now

"The best time to plant a tree was twenty years ago. The second best time is now."

Somewhere in America right now, a government executive is sitting in a conference room listening to a vendor presentation that will shape the next decade of their organization's operations. The slides are polished. The demos are seamless. The ROI projections are compelling. And the structural conditions that will determine whether this project succeeds or fails have already been set, most of them invisibly, most of them irreversibly, and none of them addressed in the presentation being delivered.

I know this because I've been in that room hundreds of times. I've been the advisor who arrived after the contracts were signed and the damage was already baked in. I've been the consultant who told leadership what they didn't want to hear: that the deal they celebrated was structured to serve the vendor's interests, not theirs. And I've been the strategist who helped organizations build the structures that prevented those outcomes in the first place, when they were willing to invest in preparation before the pressure of implementation made clear thinking impossible.

This book exists because too many government leaders are making the most consequential technology decisions of their careers without the information they need to make them well.

A CONVERGENCE OF FORCES

Three forces are converging right now that make this book not just relevant but urgent.

The first is a ticking clock. SAP, the dominant ERP platform in government for over two decades, has announced end-of-support for its legacy ERP versions. Oracle is pushing customers toward cloud migrations. Workday, Infor, and other platforms are aggressively courting government organizations with promises of modern, cloud-native experiences. The result is that thousands of government entities, including cities, counties, states, school districts, utilities, and special authorities, are simultaneously entering the market for new enterprise systems. Many of them are doing so for the first time in fifteen or twenty years, with teams that have never navigated a procurement of this scale and complexity.

The second force is artificial intelligence. AI is transforming what's possible in government technology, from how requirements are gathered and analyzed to how systems are configured, tested, and optimized. But AI is also being weaponized by vendors as a marketing tool, a way to justify premium pricing for capabilities that are poorly understood by the buyers being asked to pay for them. The gap between what AI can genuinely deliver and what vendors claim it can deliver has never been wider, and government leaders who can't distinguish between the two will pay dearly for that confusion.

The third force is the most dangerous: the accumulated weight of decades of failed implementations. Government technology projects fail at rates that would be considered scandalous in any other sector. Studies consistently show that the majority of large-scale government IT projects run over budget, behind schedule, or fail to deliver their intended benefits. The Standish Group's research has documented this pattern for thirty years. McKinsey's analysis confirms it. And yet the fundamental structures that produce these failures, including vendor-led processes, conflicted advisory relationships, inadequate preparation, and governance frameworks designed to rubber-stamp rather than challenge, remain essentially unchanged.

These three forces create a moment of extraordinary risk and extraordinary opportunity. Risk, because the sheer volume of organizations entering the market simultaneously means vendors are stretched thin, advisory talent is scarce, and the pressure to move quickly will tempt leaders to skip the preparation that separates successful implementations from expensive disasters. Opportunity, because the tools, frameworks, and governance models now available to government organizations are genuinely powerful enough to change outcomes, if leaders know they exist and have the discipline to use them.

WHY I WROTE THIS BOOK

I founded Avèro Advisors on a principle that sounds simple but is remarkably rare in the consulting industry: the people advising you on your technology decisions should have no financial relationship with the vendors competing for your business. They should succeed only when you succeed. Their incentives should be purely and permanently aligned with yours.

Over the years, I've watched this principle proven right in ways that are sometimes gratifying and often heartbreaking. Gratifying when an organization invests in Phase Zero preparation and enters procurement with the clarity and leverage that produce genuinely competitive vendor proposals. Heartbreaking when I see the aftermath of implementations where no one in the room had both the authority and the incentive to ask the hard questions, where talented, dedicated public servants were overwhelmed by vendor sales operations that had been refined across thousands of engagements.

I wrote this book because I can't be in every conference room where these decisions are being made, but the principles that protect government organizations can be. The structural insights that separate Birmingham's £216 million catastrophe from the successes at Monroe County, New York and the Port of Portland, Oregon aren't proprietary secrets. They're patterns of preparation, governance, and independence that any organization can adopt if they understand them clearly enough to act on them.

Monroe County and the Port of Portland took Phase Zero seriously. They invested in understanding their current state before vendors arrived to shape the conversation. They had IT and business leadership that understood the stakes, that recognized a technology modernization is not merely a software upgrade but a transformation that touches every department, every workflow, and every employee in the organization. Their success was not accidental. It was structural.

I also wrote this book because the emergence of AI has created a window of genuine transformation that will close. Government organizations that learn to use AI-enabled tools for requirements gathering, vendor evaluation, and implementation governance will leapfrog those that don't. But organizations that adopt AI without the governance frameworks to use it responsibly will simply add a new category of risk to an already perilous process. This book shows you how to do the former while avoiding the latter.

WHO THIS BOOK IS FOR

This book was written for government leaders who bear the weight of technology decisions that will outlast their tenure. If you are a city manager, county administrator, CIO, CFO, or department head facing an ERP modernization, a cloud migration, or a major technology procurement, this book was written for you.

It was also written for the elected officials and oversight bodies who approve the budgets and bear the political consequences when implementations go wrong. One of the persistent tragedies of government technology is that the people who authorize spending rarely have the technical context to evaluate what they're approving, and the people with technical context rarely have the authority or the incentive to share unvarnished assessments of risk.

If you're a government technology professional who has watched implementations unfold with a growing sense of unease, sensing that something was structurally wrong but unable to articulate exactly what or convince leadership to change course, this book gives you the language, the evidence, and the frameworks to make that case.

And if you're a vendor or system integrator reading this with curiosity or concern, know that this book isn't anti-vendor. It's anti-capture.

The best vendors and integrators, the ones who deliver genuine value and build lasting client relationships, thrive in environments with independent oversight and transparent governance. It's only the vendors whose business models depend on information asymmetry, scope creep, and client dependency who have reason to resist what this book advocates.

WHAT MAKES THIS BOOK DIFFERENT

The shelves are not exactly overflowing with books about government technology implementation, but the ones that exist tend to fall into two categories: academic texts that describe problems without offering actionable solutions, and vendor-sponsored guides that present their products and methodologies as the answer to every challenge. This book is neither.

Every chapter in this book is built from direct experience guiding government organizations through the decisions that determine outcomes. The case studies aren't theoretical. They're drawn from real implementations where real decisions produced real consequences for real communities. Birmingham City Council's £216 million ERP failure isn't an abstract cautionary tale; it's a forensic examination of how structural conditions made catastrophe virtually inevitable regardless of how hard anyone worked or how good their intentions were. The successes at Monroe County and the Port of Portland aren't marketing testimonials; they're detailed examinations of the specific structural choices that aligned everyone's interests with the project's success. These were organizations with IT and business leaders who understood that preparation is not a phase you rush through to get to the "real work." Preparation is the real work.

This book also confronts something most government technology guides avoid entirely: the role of AI in modern implementations. Not AI as a buzzword or a slide deck talking point, but AI as a practical tool that is already reshaping how requirements are gathered, how vendors are evaluated, and how governance is maintained throughout complex implementations. You'll learn what AI can genuinely do today, what it can't, and how to tell the difference when a vendor is presenting you with a vision of the future that may or may not have any relationship to the product they're actually selling.

Most importantly, this book is honest about the uncomfortable dynamics that govern technology procurement in the public sector. It names the conflicts of interest that permeate advisory relationships. It explains the structural incentives that drive vendor behavior. It shows how organizations get captured by the very firms they hired to help them. And it provides concrete, actionable strategies for building the independence that prevents capture from occurring.

HOW TO USE THIS BOOK

This book is organized to mirror the journey of a successful modernization, from the earliest recognition that change is necessary through the governance structures that sustain it after go-live. You can read it straight through to build a comprehensive understanding, or you can turn directly to the section that addresses your most immediate challenge.

Part I examines why government technology projects fail structurally, using Birmingham, Monroe County, and the Port of Portland as case studies that reveal the patterns separating catastrophe from success. If you need to build the case for independent oversight with your leadership team, start here.

Part II covers Phase Zero, the preparation that creates clarity before vendors arrive. This is where AI-enabled approaches, governance structures, and readiness assessments transform what has traditionally been a rushed, superficial exercise into the foundation that everything else is built on. If your organization is about to begin a modernization and you haven't yet invested in preparation, this section is your most urgent reading.

Part III addresses execution: vendor selection, contract negotiation, and governance structures that maintain independence throughout implementation. If you're already in procurement or mid-implementation and sensing that something is structurally wrong, turn here.

Part IV focuses on the human dimensions, including organizational change management, stakeholder alignment, and the advisory relationships that determine whether governance structures function as intended or become bureaucratic theater. If your implementation is technically sound but struggling with adoption and resistance, this section will show you why.

Throughout the book, you'll find frameworks, templates, and checklists that you can apply immediately. The appendix provides a complete toolkit of practical resources, and additional tools are available online at STARTATZERO.AVEROADVISORS.COM.

THE STAKES HAVE NEVER BEEN HIGHER

Let me be direct about what's at stake. The decisions you make in the next twelve to twenty-four months about your organization's technology infrastructure will determine how your government operates for the next decade or more. These aren't decisions you get to revisit. An enterprise software implementation, whether it involves your financial systems, human resources and payroll, utility billing, permitting and land management, enterprise asset management, or any combination of the mission-critical platforms that keep government running, is not a pilot program you can quietly sunset if it doesn't work out. It is a transformation that touches every department, every process, every employee, and every citizen who depends on the services your organization provides. Throughout this book, I use "ERP" as shorthand for this entire ecosystem of enterprise systems, because the principles of independent guidance apply equally whether you are replacing a single platform or modernizing your entire technology portfolio.

The urgency of getting this right has never been greater. Artificial intelligence is accelerating the gap between agencies that modernize and those that cling to legacy systems, because AI capabilities require the clean, structured, accessible data that only modern platforms provide. Organizations that act now will leverage AI to compress implementation timelines, surface requirements that manual processes miss, and deliver value in weeks rather than the years that traditional approaches demand. Organizations that wait will find themselves falling further behind, unable to take advantage of tools that are already transforming how governments procure, implement, and optimize enterprise technology.

When these implementations succeed, they transform government operations in ways that genuinely serve the public interest: faster service delivery, better resource allocation, greater transparency, and reduced costs. When they fail, the consequences extend far beyond IT departments and budget line items. They consume resources that could have funded schools, infrastructure, and public safety. They destroy the careers of dedicated public servants who were set up to fail by structural conditions they didn't create and couldn't control. And they erode public trust in government's capacity to manage complex initiatives, trust that, once lost, takes a generation to rebuild.

You don't have to be the leader who presides over the next Birmingham. You can be the leader who builds the next Monroe County or the next Port of Portland. But that choice is made in the structural decisions you make before implementation begins, not in the heroic efforts you make after things go wrong.

Turn the page. Let's start at zero.

PART I

Why Projects Fail Structurally

"Government IT disasters are not accidents, they are the predictable result of structural flaws in how projects are governed, procured, and executed. Understanding why projects fail is the first step toward building systems that succeed."

1

*The £216 Million Wake-Up Call:
Why Independent Consulting Beats
Vendor-Led Implementations*

WHAT YOU'LL LEARN

Every government leader has heard horror stories about implementations that consume budgets, destroy careers, and leave organizations worse off than before they started. However, the most revealing disasters aren't the ones caused by bad software or incompetent staff but those that happen to capable organizations making reasonable decisions—organizations that do everything they're told is right and still walk straight into catastrophe.

This chapter examines two organizations that faced identical challenges and made opposite choices. Birmingham City Council, Europe's largest local authority, transformed a working system into a £216 million disaster that bankrupted the city and ended careers. In contrast, Monroe County, New York, facing the same aging infrastructure and modernization pressures, is delivering a successful implementation that's proceeding on schedule and on budget. The difference wasn't the software they chose, the vendors they hired, or the expertise of their staff. The key distinction was whether someone in the room had both the authority and the incentive to protect their interests when vendor's did not align.

Understanding what separated these outcomes establishes the foundation for everything that follows in this book:

- The Phase Zero preparation that creates clarity before vendors arrive
- The procurement strategies that attract capable partners while maintaining independence
- The governance structures that keep implementations on track
- The organizational change management that determines whether new systems transform operations or simply replace one set of problems with another

THE SUCCESS STORY NOBODY REMEMBERS

Before examining the disaster that made headlines worldwide, we need to understand the success that makes Birmingham's failure so tragic and so preventable.

For 20 years, Birmingham City Council operated quietly one of the world's most successful government Enterprise Resource Planning (ERP) implementations, a distinction that deserves emphasis because success in government technology is rare enough to warrant celebration. Their SAP system, launched in 1999, came in on time and on budget, which is a phrase rarely used in government IT that it bears repeating: *on time and on budget.* The system delivered £100 million in savings over its first 5 years, while boosting productivity by 15% across the council's operations.[1] Birmingham City University was so impressed by the implementation outcomes that they partnered with SAP to create an entire master's degree program built around studying what Birmingham had accomplished.[2]

For two decades, that SAP system performed the unglamorous work of local government—processing payroll for 13,000 employees, cutting checks to suppliers, collecting council tax from residents, and managing budgets across dozens of departments. The system wasn't exciting or transformational in the buzzword sense that vendors love to promote, and it never generated headlines. It simply worked, reliably and consistently, enabling Birmingham to serve its citizens, while other councils struggled with failed implementations and technology crises.

Understanding this history matters because it reveals the magnitude of what Birmingham subsequently destroyed and the preventability of the disaster that followed.

FROM SUCCESS TO CATASTROPHE

In 2018, Birmingham faced a new challenge: SAP had announced end-of-support for older ERP versions by 2025, requiring customers

---

1   https://www.computerweekly.com/news/2240048248/Birmingham-City-Council-completes-SAP-rollout

2   https://www.ukauthority.com/articles/auditor-s-report-highlights-failings-in-birmingham-city-council-erp-programme/

to either upgrade within the SAP ecosystem or migrate to alternative platforms.[3] The existing system, managed through a joint venture with Capita, called Service Birmingham, incurred annual costs of approximately £5.1 million. Council leadership determined they needed to evaluate options for moving to a modern platform, either a newer SAP version or an alternative provider—a reasonable decision that organizations across the globe were making and continue to make today.

Birmingham selected Oracle's Cloud ERP platform as their replacement, and this choice was entirely defensible. Oracle offers a robust suite of ERP and HCM solutions that have been successfully deployed at government organizations worldwide, with references and case studies demonstrating the platform's capability for public sector operations. So, the software or the vendor was not the problem.

The problem was everything Birmingham didn't do.

Birmingham didn't invest in Phase Zero preparation to understand their current state, document their actual requirements, and establish clarity about what success would look like before vendors began shaping the conversation. They didn't establish independent oversight to provide a perspective in the room whose sole purpose was protecting council interests. They also didn't create a governance structure that separated advice from delivery and could evaluate vendor recommendations objectively rather than accepting them as directives. Instead, they handed their transformation entirely to vendors and system integrators who had everything to gain from

---

3   https://www.computerweekly.com/news/366538632/Birmingham-council-leader-admits-100m-budget-hole-in-ERP-upgrade

expanded scope, extended timelines, and additional complexity that might or might not serve Birmingham's actual needs.[4]

Consider the structural problem that was created: Birmingham effectively hired the architect, contractor, and building inspector from the same company, or from companies whose financial ties bound their interests together. There was no one in the room with both the power and the motive to say that the recommendation served the vendor not the council, that the scope wasn't needed, or that the timeline extension reflected vendor staffing challenges rather than project requirements.

By September 2025, the consequences of this structural failure had become undeniable. Birmingham City Council Leader John Cotton convened an emergency meeting that would echo through British local government for years, confronting a situation that had spiraled far beyond anyone's initial projections.[5] The £19 million project had consumed over £100 million with no clear endpoint in sight.[6] The council faced a £750 million budget shortfall driven significantly by technology costs. Birmingham, Europe's largest local authority serving over a million citizens, was effectively bankrupt.[7]

The full scope of the disaster defied belief: total costs exceeding £216 million; a timeline stretching past 7 years with basic functionality still missing; systems lacking audit trails that any government finance operation requires; and human consequences including 600

---

4   https://www.theregister.com/2023/09/18/birmingham_city_council_oracle/

5   https://www.bbc.com/news/uk-england-birmingham-66716452

6   https://www.computerweekly.com/news/366538632/Birmingham-council-leader-admits-100m-budget-hole-in-ERP-upgrade

7   https://www.theguardian.com/uk-news/2023/sep/05/birmingham-city-council-financial-distress-budget-bankruptcy

employees laid off, services slashed across the council, and libraries closed to fund a technology project that still couldn't perform basic statutory reporting.[8]

WHY VENDOR-LED IMPLEMENTATIONS FAIL PREDICTABLY

Birmingham's story demands attention not because it represents an unusual catastrophe but because it is a typical one. The same pattern has played out with SAP, Tyler, Workday, Microsoft, and Oracle implementations across the globe, with the specific vendor mattering far less than the structural conditions that allowed failure to compound unchecked.

A senior finance manager who lived through Birmingham's catastrophe described the experience in terms that illuminate why these failures follow such predictable patterns:

> "We kept waiting for someone to protect us, to push back on the vendor's recommendations when they didn't seem right. But everyone in the room was incentivized for the project to grow. The vendors, the integrators, even the consultants we hired for advice, they all made more money when things got complicated. We never stood a chance."[9]

8   https://www.computerweekly.com/news/366615147/Birmingham-city-council-ERP-rebuild-to-hit-131m-with-more-modules-delayed-to-september-2025

9   https://averoadvisors.com/case-studies/monroe-county

The following three structural flaws consistently doom vendor-led implementations, regardless of the software platform or vendors involved.

The first flaw involves fundamental incentive misalignment that creates conflicts built into the project's foundation. When organizations hire system integrators on time-and-materials contracts, the integrator's revenue correlates directly with project duration and complexity. Extended timelines generate more billable hours. Expanded scope triggers change orders that increase total contract value. Efficiency becomes a financial penalty for the integrator because finishing 6 months early means losing 6 months of revenue. Meanwhile, the client's incentive points in exactly the opposite direction—finishing as quickly as possible with minimal scope expansion. When the entity responsible for managing schedule and scope profits from both slipping, the conflict of interest is mathematical rather than moral, structural rather than personal. Good people operating within bad structures produce bad outcomes.

The second flaw involves expertise assumptions that prove unfounded when implementations encounter government-specific complexity. Software vendors bring deep technical knowledge of their platforms, and system integrators bring implementation methodology refined across hundreds of projects. But neither typically brings genuine understanding of government operational requirements, the statutory obligations, political dynamics, citizen service expectations, and accountability structures that make public sector implementations fundamentally different from private sector deployments. Birmingham's auditors found that implementation teams fundamentally misunderstood UK local government requirements, delivering a system that couldn't perform basic statutory reporting that any council finance operation requires.[10] Technical excellence means nothing when the system can't do what government actually needs.

The third flaw involves accountability vacuums that prevent anyone from owning outcomes when implementations go wrong. When projects fail, the finger-pointing begins immediately and continues indefinitely. Software vendors blame system integrators for poor configuration and inadequate change management. System integrators blame software limitations and unexpected platform behaviors. Everyone blames "changing requirements" that supposedly emerged from the client, as if organizations shouldn't be allowed to clarify their needs as understanding develops. Nobody accepts responsibility for outcomes because the structure allows everyone to point elsewhere.

THE MONROE COUNTY MODEL: HOW INDEPENDENCE CHANGES EVERYTHING

While Birmingham's disaster was unfolding across the Atlantic, Monroe County in New York faced the exact same challenge that had initiated Birmingham's catastrophe—an aging SAP system requiring modernization, end-of-support deadlines approaching, and pressure to move toward modern cloud-based platforms. The parallels were striking enough that Birmingham's experience could have served as

---

10   https://averoadvisors.com/case-studies/monroe-county

a warning, but only if Monroe County structured their approach to avoid the same traps.

The County issued an RFP for consulting services to guide their modernization journey, and multiple firms responded with proposals, including several "Big Four" firms whose scale and brand recognition typically dominate such competitions. The presentations that followed revealed fundamentally different philosophies about what government technology consulting should accomplish.

The large firms arrived with familiar scripts emphasizing their partnerships with major software vendors, their proprietary accelerators and methodologies, and their track records implementing systems at scale. When County leadership asked about Birmingham's experience and what it might mean for Monroe County's approach, the responses were dismissive: "Birmingham's situation is unique," implying that whatever had gone wrong there couldn't happen here.

When my firm's turn came, we started with questions rather than answers, seeking to understand Monroe County's specific situation rather than immediately proposing solutions. Then we said something that made the room visibly uncomfortable:

> "Monroe County could end up exactly like Birmingham if you make the same fundamental mistake they made. Birmingham jumped straight into vendor selection without proper preparation, without understanding their current state clearly enough to evaluate what vendors were proposing. More critically, they had no independent voice in the room whose sole purpose was protecting their interests when vendor

recommendations might serve vendor interests more than council interests."

The County's evaluation committee asked the question that would determine everything that followed: "So what exactly would your role be if we selected you?"

> "We would serve as your independent PMO and oversight partner throughout the entire journey," we explained. "When you eventually select your software vendor, we'll help evaluate them objectively because we don't receive commissions, referral fees, or partnership revenues from any of them. When your system integrator proposes scope changes or timeline extensions, we'll assess whether those proposals serve your interests or theirs. We're like having your own expert representation in every room where decisions get made."[11]

Monroe County selected our firm, and we spent 18 months in Phase Zero preparation before the County ever issued an RFP for software.[12] During that period, we documented current state operations in detail that would later enable objective evaluation of vendor proposals. The specific pain points and requirements that any new system would need to address were also identified. Governance structures that would maintain oversight throughout implementation

---

11   https://averoadvisors.com/case-studies/monroe-county
12   https://www.panorama-consulting.com/resource-center/erp-report/

were established. Finally, we prepared the organization for the change that was coming rather than allowing vendors to shape expectations based on what they wanted to sell.

When Monroe County finally went to market for software, they possessed what Birmingham had lacked: clarity. The RFP didn't ask vendors for vague promises about "transformation" or "best practices." It posed specific scenarios drawn from actual Monroe County operations, requiring vendors to demonstrate how their platforms would handle the County's real complexity rather than idealized use cases that any system could address.

Six vendors responded to that RFP. After rigorous evaluation with our firm providing independent assessment unmarried to any vendor relationship, Monroe County selected Workday as their software platform and AVAAP as their system integrator. But the structure Monroe County created diverged completely from Birmingham's approach by maintaining three distinct layers with clearly separated responsibilities (Figure 1.1).

The foundation layer consists of the software vendor (in this case Workday) whose responsibility centers on product development, platform stability, and feature delivery. The execution layer consists of the system integrator (i.e., AVAAP) whose responsibility encompasses configuration, customization where necessary, and technical implementation. The oversight layer consists of the independent PMO, our firm, whose responsibility focuses on scope control, change management, schedule integrity, and client advocacy. Above all three layers sits the Executive Sponsor, the County executive empowered to make final decisions when the layers disagree, ensuring that someone with authority and accountability owns outcomes rather than allowing disputes to drift unresolved.[13]

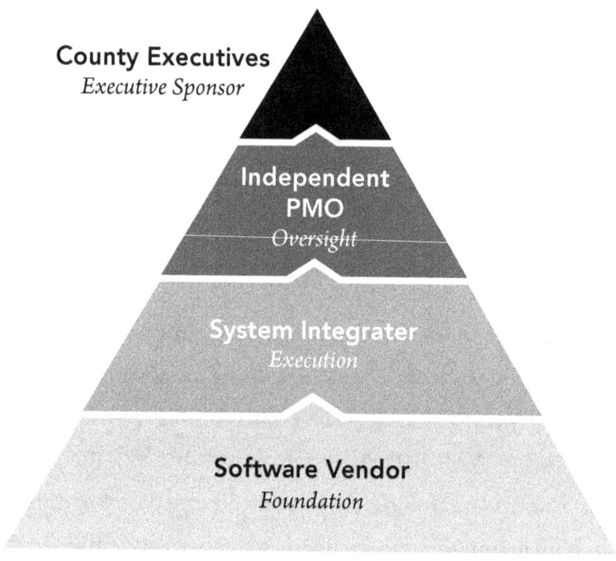

Figure 1.1 The Monroe County approach: Three distinct layers with clearly separated responsibilities.

This structure created something Birmingham never had—productive tension between parties whose interests don't automatically align.[14] When a vendor suggested additional modules that would expand the implementation footprint, we asked whether those modules were necessary for go-live success or represented scope creep

---

13   https://www.panorama-consulting.com/resource-center/erp-report/

14   https://www.ukauthority.com/articles/auditor-s-report-highlights-failings-in-birmingham-city-council-erp-programme/

that could be deferred to future phases. When the system integrator proposed timeline extensions, we pressed for specific issues driving the delay. In addition, the integrator was to resolve internally whether those issues reflected genuine complexity or resource management challenges. When configuration decisions arose defining long-term operations, we ensured County staff understood the implications rather than deferring to vendor recommendations without scrutiny.

A County leader described the difference this structure created:

> "Having independent oversight in the room changed everything about how decisions got made. Suddenly, vendor recommendations became discussions rather than directives. We had someone whose only job was protecting our interests, someone who could ask the uncomfortable questions without worrying about damaging a vendor relationship they needed for future business. It's like having your own attorney when buying a house. You wouldn't sign contracts involving millions of dollars without independent representation, so why would you implement an ERP system that way?"

RECOGNIZING THE WARNING SIGNS

Organizations heading toward Birmingham-style disasters rarely recognize the trajectory until crisis becomes undeniable. The warning signs appear early, but they're easy to rationalize when projects maintain momentum and vendors provide reassuring status reports.

Understanding these patterns opens the door to intervention, while course correction remains possible.

During procurement, several warning signs tend to appear:

- The absence of Phase Zero preparation work that establishes clarity before vendors shape the conversation
- Lack of independent oversight structure that could evaluate vendor proposals objectively
- Vendor-led requirements gathering that allows vendors to define the problems they're proposing to solve
- Compressed timelines justified by urgency to "maintain momentum" rather than adequate time for proper evaluation
- Decision committees where vendor representatives outnumber or dominate client voices

During implementation, warning signs include:

- Patterns where every discovery session leads to scope expansion rather than occasional scope refinement balanced by scope reductions elsewhere
- Mounting change orders treated as normal project evolution rather than signals of inadequate initial planning
- Original consultants who impressed during sales being replaced by junior resources who lack the expertise that justified selection
- Testing phases compressed to meet deadlines rather than timelines adjusted to accommodate thorough testing

- No individuals who could say no to vendor expansion recommendations regardless of their merit.

During emerging crisis, warning signs are:

- Sudden engagement of vendor executives who had been absent during routine operations
- Multiplication of "tiger teams" and "war rooms" that signal problems too severe for normal project management
- Lawyers entering conversations that had previously been technical and operational
- Media inquiries about project status that suggest external visibility into internal problems
- Careers beginning to end as accountability finally arrives too late to prevent disaster.

Ask yourself a simple question right now: Who in your current or planned implementation has both the authority and the incentive to say no to vendor expansion when vendor interests diverge from yours? If you cannot name that person, if no one in your structure occupies that role, then you don't have independent protection, and you're relying on vendor goodwill to substitute for structural safeguards. Birmingham relied on that same hope, and £216 million later, they learned what that hope was worth.

THE PATH FORWARD

As you read this, your organization likely faces pressures similar to those experienced by Birmingham and Monroe County. Legacy systems are aging, and vendors are announcing end-of-support dates that create urgency. Technical debt accumulates as workarounds, and patches maintain systems never designed for current demands. Citizens expect modern services delivered through digital channels your current infrastructure struggles to support. Federal and state mandates impose requirements your legacy systems cannot meet without expensive customization.

The vendors are already calling, and their presentations are impressive. They're promising transformation, innovation, AI, and revolution. They have case studies, references, and demonstrations that make everything seem achievable. And there's nothing inherently wrong with their software or their implementation expertise, both of which can deliver genuine value when properly managed and overseen.

What vendors cannot provide is someone whose sole purpose is protecting your interests. They cannot provide the productive tension that emerges when independent oversight questions recommendations that might serve vendor interests more than the client's. They cannot provide the accountability structure that prevents problems from hiding behind optimistic status reports until crisis makes them undeniable.

Birmingham had a working system that served the council effectively for 20 years, and they destroyed it. Not because they chose the wrong software, not because Oracle's platform couldn't meet their needs, not because their staff lacked capability or commitment. They destroyed it because they had no one protecting their interests when vendor interests diverged, no structure that could say no when it was the right answer, and no independence that could have maintained clarity when complexity and cost began compounding beyond reason.

Monroe County is succeeding with the same fundamental challenge, modernization pressures, and vendor landscape, simply because they built a structure that aligned everyone's interests with their own success. Independent oversight doesn't guarantee success, but its absence virtually guarantees the conditions where failure becomes likely and where early warning signs get rationalized until intervention becomes impossible.

▌KEY TAKEAWAYS

- **Software doesn't fail; governance structures do.** Birmingham's disaster wasn't about Oracle specifically, and substituting any other vendor's name would change nothing about the underlying dynamics. The same failure pattern occurs with every software platform when independent oversight is absent, when no one in the room has both the authority and the incentive to protect client interests, and when vendors face no productive tension that challenges recommendations serving their interests over yours. Technology matters far less than the structure within which technology decisions are made.

- **Phase Zero preparation determines implementation success long before implementation begins.** The 18 months Monroe County invested in understanding their current state, documenting their actual requirements, establishing governance structures, and preparing their organization for change will save them years of crisis management. Vendors who urge you to skip this preparation, promise their methodologies will surface requirements during implementation, and characterize Phase Zero as unnecessary delay are asking you to make the same mistake that cost Birmingham £216 million and counting.

- **Independent oversight isn't optional; it's the structural foundation that makes everything else possible.** You wouldn't purchase a home without your own inspector evaluating the property, sign complex contracts without your own attorney reviewing the terms, or undergo major surgery without seeking a second opinion from a physician whose income doesn't depend on performing the procedure. Why would you implement an enterprise system that will shape your operations for the next decade without independent representation protecting your interests throughout the journey?

▌ YOUR IMMEDIATE ACTION

Within the next 2 weeks, conduct an honest assessment of your current or planned implementation's structural integrity by answering three questions that reveal whether you have genuine protection or merely the appearance of oversight:Identify everyone on your implementation team, both current staff and planned participants, who profits financially if scope expands, timelines extend, or complexity increases. This includes vendors and system integrators billing time and materials, consultants with partnership arrangements or referral relationships with vendors, and anyone else whose compensation correlates with project growth rather than project success. Understanding who benefits from expansion reveals whose recommendations require independent validation.

1. Identify who in your structure has genuine authority to say no to vendor recommendations—someone who can reject proposed scope changes, push back on timeline extensions, and question whether vendor suggestions serve your interests or theirs. This person must have both the positional authority to make their "no" stick and the independence from vendor relationships that makes their judgment trustworthy. If this person doesn't exist, you have a structural gap that no amount of good intentions can fill.

2. Identify who is accountable solely to you—someone whose only success metric is your success, who receives no revenue from software vendors or system integrators, and whose professional reputation depends entirely on your implementation achieving its objectives. This is the role that Monroe County filled with independent oversight and that Birmingham never established.

If you cannot answer these questions, or if the answers reveal gaps in your protection structure, you're on a path that leads toward Birmingham rather than Monroe County. But unlike Birmingham at this stage in their journey, you still have time to change the course. The chapters that follow will show you exactly how to do that, beginning with the Phase Zero preparation that creates clarity before vendors arrive and continuing through procurement, vendor selection, governance, and organizational change management that maintain independence throughout implementation.

The choice between Birmingham's and Monroe County's trajectory isn't made once in a dramatic decision point. It stems from dozens of structural choices about who participates in decisions, whose interests get represented, and whether anyone in the room has both the authority and the incentive to protect you when protection is needed most.

2

Legacy Time Bombs
Why Waiting Costs More Than Acting

WHAT YOU'LL LEARN

- Why each month of delay multiplies your risk exponentially
- The four risk multipliers that compound over time
- How to calculate the true cost of postponement
- When "wait and see" becomes "too late to save"

THE CALL THAT CHANGES EVERYTHING

Imagine the following scenario: It's 3:47 AM when your phone rings. In that disorienting moment between sleep and consciousness, you already know that nobody calls at that time with good news.

"The payroll system is down." Your CIO's voice is steady, professional, but you hear the edge underneath, the slight tremor that says this isn't just another midnight server restart. "Completely down. We can't process payroll tomorrow."

You're fully awake now, sitting up in bed, your spouse stirring beside you.

"When can we get it back up?"

There is a pause—that particular kind of silence that tells you everything before the words come. "We're not sure we can. The database is corrupted, and it's been running on a server we can't get parts for anymore. The last person who really understood the architecture retired 18 months ago. We convinced him to document what he could before he left, but . . ." Yet another pause. "The vendor hasn't supported this version since 2015. We're trying to find someone, anyone, who remembers how this works."

This is how legacy system failure announces itself—not with months of graceful degradation and warning signals you can plan around, but with a phone call that turns your entire organization upside down before you've had your first cup of coffee. By 7 AM, you'll be on a conference call explaining to union representatives why 5,000 employees won't be getting their direct deposits. By noon, the media will be calling, and some resourceful reporter will have already submitted a public records request for all emails related to "system modernization delays." By evening, the mayor will want answers you don't have, and the City Council will be scheduling emergency hearings where you'll explain why the system you've been calling stable for the past 5 years just destabilized everything.

This scenario isn't fiction. It's played out with variations in Rochester City Schools, where nearly 1,000 employees—many living paycheck to paycheck—went unpaid when the new payroll system failed after an update.[15] It also happened at the Federal Aviation Administration (FAA), where a corrupted database file in a decades-old messaging system brought American air traffic to a complete standstill, grounding thousands of flights and stranding millions of passengers because a critical system that never failed finally did.[16] It even may be happening right now in your organization, where the comfortable fiction that "still working" equals "safe to keep running" is about to meet the uncomfortable reality of catastrophic failure.

THE COMFORTABLE LIE OF "IF IT AIN'T BROKE..."

There's a seductive logic to leaving legacy systems alone, a siren song that every government executive has heard and most have hummed along with. The melody goes something like this: *These systems work. They're paid for. People know how to use them. The risk of change feels greater than the risk of standing still.*

Your Finance Director, a sharp professional who's shepherded the organization through multiple budget crises, loves to remind everyone at leadership meetings: "This system has run perfectly for 15 years. It's processed over 400 payroll cycles without a major incident. Why would we spend the millions we need for services, for staff, for actual

---

15 Justin Murphy, "RCSD Payroll Problems Persist After Fix Fails for Hundreds of Employees," *Rochester Democrat and Chronicle*, August 29, 2025.

16 FAA News, "FAA Statements on NOTAM System Outage," *Federal Aviation Administration*, January 11, 2023.

citizens' needs to replace something that works? Show me the ROI on fixing something that isn't broken."

Your IT Director, equally sharp but bearing the exhaustion of someone who's been issuing warnings that fall on deaf ears, quietly responds: "It's not broken yet. But when it breaks, we won't be able to fix it. We're one hardware failure away from disaster. One ransomware attack from paralysis. One retirement from losing irreplaceable knowledge."

This conversation, with minor variations, happens in every government agency running legacy systems. It happened in Birmingham before their implementation disaster. It happened in Rochester before their payroll catastrophe. It probably happened in your organization, maybe last week, maybe this morning.

Usually, the voice of caution loses this argument. But here's what that well-intentioned Finance Director doesn't understand: Legacy systems aren't like wine, improving with age. They're like bridges, accumulating stress fractures with each passing day. Every day they continue working reinforces the illusion of stability, right up until the moment of catastrophic failure.

THE FOUR HORSEMEN OF LEGACY APOCALYPSE

After analyzing hundreds of government system failures, I've identified the following four risk multipliers that compound over time, turning manageable transitions into existential crises.

1. The Knowledge Extinction Event

Consider an archetype you'll find in nearly every government agency. Let's call him "Bob." Bob is 63 years old and has worked for your organization for 31 years. He wrote the integration software between your property tax system and the general ledger, the one with undocumented patches that accumulated over decades like barnacles on a ship's hull. Bob knows why the system throws that cryptic "Error 447-B" every third Tuesday. When something breaks, Bob fixes it, not because there's a manual, but because he was there when it was built.

Bob is retiring in 18 months, and when Bob walks out the door, 20 years of undocumented system knowledge goes with him. The cryptic error messages that Bob can interpret in seconds will become multiday mysteries.

This isn't just a story. In April 2020, the State of New Jersey's 40-year-old COBOL-based unemployment system buckled under the surge of pandemic-related claims. Governor Phil Murphy made a national public plea on television, desperately searching for COBOL programmers to come out of retirement to help. They were facing a catastrophic failure of a critical public service, not because of a lack of funding at that moment, but because of a decade-long failure to address the predictable extinction of critical knowledge.[17]

2. The Security Time Bomb

On May 7, 2021, the Colonial Pipeline, which supplies nearly half the fuel for the U.S. East Coast, was shut down by a ransomware attack.

---

17   Alfred Ng, "'We Need Help': States Desperate for COBOL Programmers to Help With Aging Unemployment Systems," *CNBC*, April 4, 2020.

The entry point was a single compromised password for a legacy VPN account that was not protected by multifactor authentication.[18] What chance does a local government running 20-year-old systems have against modern ransomware? The answer is brutal: None.

Legacy systems are unlocked doors. They predate modern encryption and zero-trust architecture (technical speak for modern security techniques and protocols). Worse, cyber insurance carriers are increasingly refusing to cover damages from attacks on unsupported, end-of-life systems. They've done the math and know these systems are indefensible. If you're running unsupported software and get hit, you're on your own.

3. The Compliance Avalanche

Modern systems adapt to new compliance requirements through updates. Legacy systems require custom code, manual processes, and expensive third-party tools, if compliance is even achievable at all.

When the IRS introduced new tax reporting requirements, a state revenue department I worked with discovered its 15-year-old system couldn't generate the data in the required format. Their solution? *Forty-two employees manually reformatting data in Excel every single month.* This is a true story. The annual labor cost is over $3 million, and the modern system they delayed "to save money" would have cost $8 million, one time. They've now spent over $12 million on the manual workaround, with no end in sight. The cost of inaction has already doubled the cost of action.

---

18   William Turton and Kartikay Mehrotra, "Hackers Breached Colonial Pipeline Using Compromised Password," *Bloomberg,* June 4, 2021.

4. The Compound Interest of Technical Debt

Technical debt is like credit card debt—ignore it, and it compounds until it consumes everything. Consider the scenario presented in Figure 2.1, which has played out in dozens of governments I've evaluated.

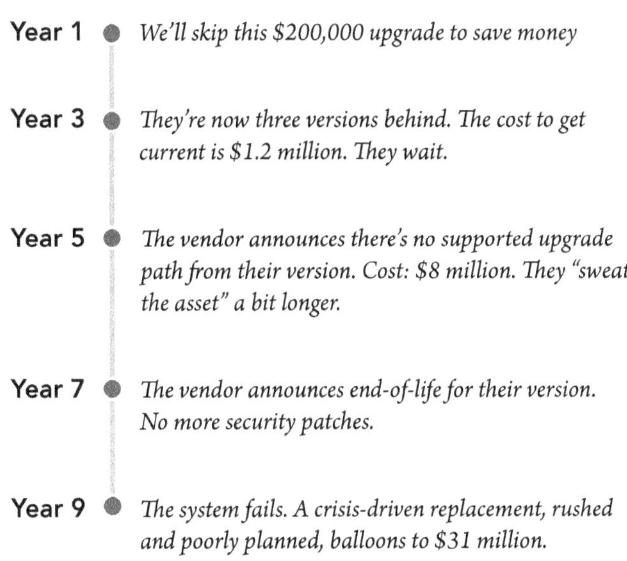

Figure 2.1 The escalating costs of
ignoring technical debt

In the graphic above, this County saved $200,000 in Year 1 to eventually spend $31 million—a 155x return on their savings.

THE WINDOW IS CLOSING

As I write this in December 2025, several forces are converging. According to the National Association of State CIOs (NASCIO), the top issue facing government IT leaders is the aging workforce and the impending retirement of staff with legacy skills.[19] Meanwhile, ransomware attacks on government agencies have been escalating, with the FBI reporting a significant increase in incidents targeting the public sector.[20]

THE PORT OF PORTLAND'S PROACTIVE PATH

While many wait for a crisis, others act. In 2024, the Port of Portland, facing aging systems that still functioned but showed increasing strain, made the crucial decision to invest in comprehensive Phase Zero preparation *before* their systems failed. They partnered with Avèro Advisors to conduct the deep analysis most organizations skip.[21] The Port's leadership recognized that "working" isn't the same as "sustainable." This is what proactive modernization looks like—careful preparation while you still have options.

---

19 NASCIO, "The State CIO Top 10 for 2024," *National Association of State Chief Information Officers*, October 2023.

20 FBI Internet Crime Complaint Center, "2023 Internet Crime Report," *Federal Bureau of Investigation*, 2024.[7] Avèro Advisors, "Port of Portland, OR Case Study," *averoadvisors.com*, Accessed September 2025.

21 Avèro Advisors, "Port of Portland, OR Case Study," *averoadvisors.com*, Accessed September 2025.

▌ YOUR IMMEDIATE ACTION

Within the next 48 hours, conduct your own Legacy Risk Assessment: List every system over 7 years old, note vendor support end dates, and identify the "Bobs" in your organization. If your "Point of No Return"—when an upgrade becomes a full replacement, or when your last expert retires—is within 24 months, you must start acting today. Because that 3:47 AM phone call doesn't care about your schedule.

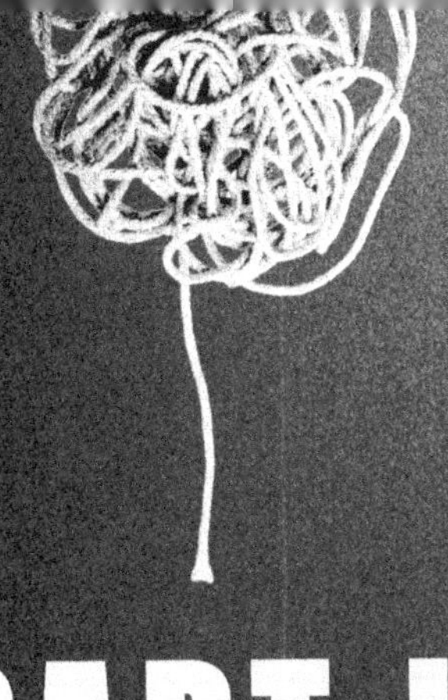

PART II

Regaining Command

Modernization is not a technology project; it is an exercise in sovereignty. Success is determined by the clarity established in Phase Zero; the quiet work of documenting reality with AI-enabled precision and building an architecture of control long before the pressure of implementation begins. To skip this investment in understanding is to surrender your digital future to the very structural flaws that lead to catastrophe.

3

The Architecture of Control
The Four Pillars of Project Sovereignty

WHAT YOU'LL LEARN

- The four structural requirements that ensure you, and not the vendor, remain in the driver's seat
- Why standard vendor contracts are architected to erode your authority
- How to build a governance structure that filters out noise and forces objective truth
- The difference between renting expertise and owning the outcome through independent, unbiased consulting teams

In the wake of Birmingham's £216 million catastrophe, the central question every government leader should ask is, *How did this happen?* How did an organization that successfully ran a complex ERP for two decades walk into such an obvious trap?

The answer isn't about technology. Oracle's software, like SAP's, is a powerful tool used successfully by thousands of organizations. The failure wasn't in the software but in the support structure built or rather, not built around the project. Birmingham failed because it lacked the foundational pillars that Monroe County instinctively understood were non-negotiable.

Successful government modernizations aren't built on software. They're built on the following four pillars that guarantee the project serves the public interest, not the vendor's profit motives.

PILLAR 1: PURE ALIGNMENT (THE "BULLDOG" PRINCIPLE)

In government ERP implementations, incentives are structurally misaligned. The System Integrator's revenue model creates inherent tension: While extended timelines don't automatically translate to higher fees, they certainly can when change control processes aren't rigorously enforced. Scope creep, unplanned customizations, and poorly managed requirements all become billable opportunities. Meanwhile, although system integrators (SIs) have to maintain formal partnerships with software vendors such as Oracle, SAP, or Workday—relationships that can ensure quality standards and technical support—the SI ultimately controls project planning, management, and execution. When that control is exercised poorly, timelines stretch, quality erodes, and projects spiral regardless of the major software brand on the letterhead.

This creates an uncomfortable dynamic: The party responsible for managing scope and change control is often the same party positioned to benefit when those processes break down.

To avoid these structural misalignments, you need an independent consultant—an advisor with no vendor affiliations or SI partnerships—whose only incentive is your success. Independent consultants have billable incentives, too, but theirs run in the opposite direction. Their continued engagement depends on project momentum and client satisfaction. When milestones slip or frustration mounts, their relationship is at risk. This creates natural alignment: The consultant succeeds when you succeed, having every reason to flag problems early, challenge unnecessary scope, and keep the vendor teams accountable.

An independent consultant has only one metric for success: yours. Their sole purpose is to ensure your project is delivered on time and on budget, providing the value promised to stakeholders.

A software vendor or their system integrator partner operates under a fundamentally different incentive structure. Their business model often depends on maximizing revenue from your account, not just initial license sales and implementation fees, but a long tail of change orders, scope expansions, and post implementation services. The data on ERP project success consistently shows that clear objectives and controlled scope are among the most critical success factors. A landmark McKinsey & Company study found that "clear objectives and scope" was the top driver of project success.[22] Yet a vendor's incentive to expand scope stands in direct tension with your need to contain it.

This isn't theoretical; it's structural. Consider a simple scenario: You discover a new requirement mid-project. An independent consultant, aligned with your budget and timeline, will first ask whether

---

22   Michael Bloch, Sven Blumberg, and Jürgen Laartz, "Delivering large-scale IT projects on time, on budget, and on value," *McKinsey & Company*, October 2012.

it's essential for go-live, whether existing functionality can address the need, or whether it can be deferred to a future phase to protect the core project. A system integrator operating under a time-and-materials contract faces a different calculus—a change order means additional billable hours and increased revenue. The incentive is to say yes and expand the engagement.

Call it the "Bulldog Principle." An independent consultant can push back on vendors, challenge questionable change orders, and hold system integrators accountable in ways that internal staff often cannot. Government project managers are juggling dozens of priorities and maintaining relationships they'll need long after go-live. An independent consultant can absorb the friction, asking the uncomfortable questions, demanding documentation, and refusing to let issues slide without jeopardizing those long-term relationships. They can be the bulldog so you don't have to be one.

This is the fundamental flaw of vendor-led oversight. The entity responsible for protecting your scope and budget is the same one that profits when they grow. Without an independent partner whose only product is your success, you enter every negotiation at a structural disadvantage.

PILLAR 2: UNBIASED SUPPORT (THE "EXPERT WITNESS" PRINCIPLE)

Imagine a legal case where your lawyer was also on the payroll of opposing counsel. You would never agree to it. Yet governments routinely accept this exact conflict of interest in their most expensive and complex projects.

When a software vendor offers "project management" or "quality assurance" as part of their implementation package, they are not providing independent oversight. They are providing self-oversight, an illusion of accountability dressed up as a service.

Industry data confirms what my experience suggests: Unnecessary change orders and unexpected scope creep are primary drivers of IT project overruns. The Project Management Institute reports that 52% of projects experience scope creep,[23] which is not accidental. For many implementers, it's a core part of the business model. With unscrupulous vendors, initial bids can be deceptively low because the profit margin, and then some will be recovered through the inevitable change orders that arise in any complex project.

An independent consultant serves as your expert witness, shielding you from this dynamic. They vet every change order, distinguishing what's truly necessary from what's merely convenient, and challenging cost estimates that appear inflated. They provide unbiased technology advice, where a vendor will reflexively solve problems with their own tools, an independent consultant can identify when a third-party solution is better and cheaper, or when a simple process change eliminates the need for costly customization altogether. And when testing reveals a flaw, they ensure the vendor fixes it under the terms of the contract, rather than allowing it to be reclassified as a new requirement necessitating a paid change order.

Without this unbiased shield, you are asking the fox to guard the henhouse and acting surprised when feathers fly.

---

23   Project Management Institute, "Pulse of the Profession 2018: Success in Disruptive Times," *pmi.org*, 2018.

PILLAR 3: GOVERNMENT EXPERTISE (THE CONTEXT PRINCIPLE)

A common mistake is hiring a large, brand-name corporate consulting firm to oversee a government project. These firms may have deep technical expertise, but technical expertise and government expertise are entirely different disciplines.

Government is not simply corporate with more bureaucracy. It operates under a unique set of constraints and rules that have no parallel in the private sector. The U.S. Government Accountability Office has repeatedly cited failure to understand the unique operational environment of government as a key factor in major IT disasters.[24]

A consultant without deep public sector experience can inadvertently lead you toward failure. They may not grasp the complexities of fund accounting, including restricted funds, grants, and enterprise funds, that determine whether your financial system can pass a government audit. They may design an HR system based on corporate best practices that collapses once it encounters union contracts or civil service laws. They may recommend procurement strategies or contract language that inadvertently violate state statutes or federal purchasing regulations. They may architect data systems without considering the legal requirements for public records retention and disclosure that govern every government agency.

Monroe County succeeded in part because their independent consultants were government specialists. They understood the unique world of public finance, public sector HR, and government procure-

---

24   United States Government Accountability Office, "Information Technology: A Framework for Assessing and Improving Enterprise Architecture Management (Version 2.0)," *GAO-10-846G*, September 2010.

ment. Birmingham's advisors, by contrast, were technical specialists who lacked the context to challenge vendor recommendations that were incompatible with a public sector environment. True expertise isn't just about knowing the software but about knowing the world in which that software must operate.

PILLAR 4: CONTINUITY (THE INSTITUTIONAL MEMORY PRINCIPLE)

Major modernizations are not short-term projects. They are multiyear journeys that begin long before a vendor is selected and continue long after the system goes live. A fatal flaw in many projects is the revolving door of consultants and project managers.

When your vendor or SI's project manager changes every 6 months, institutional knowledge evaporates, trust erodes, and momentum stalls. A vendor's project team is bound by the vendor's corporate needs, and your project manager could be reassigned to a larger, more profitable client at any moment. Your project's success is their priority only until something more lucrative comes along.

An independent firm provides a consistent team that stays with you for the entire journey. During Phase Zero, they help you define your vision and requirements, ensuring the foundation is solid. During procurement, the same team that built the requirements helps you evaluate vendors, ensuring nothing is lost in translation. During implementation, the same team provides oversight, holding the vendor accountable to the promises made during the sales process. And after go-live, the same team helps you manage the system, plan for future phases, and ensure you realize the long-term value of your investment.

This continuity is the antidote to the knowledge extinction event discussed in Chapter 2. While your internal experts are retiring and taking decades of institutional knowledge with them, an independent consultant becomes the bridge. Independent consultants serve as a stable repository of project history, decisions, and strategic intent, ensuring the vision defined in Year 1 is the reality delivered in Year 3.

These four pillars are the invisible structure that holds a successful project together. They are what makes the difference between Monroe County's quiet competence and Birmingham's public failure.

Understanding why you need an independent consultant is only half the equation. The other half is knowing when to bring them in.

Too many government agencies wait until problems emerge mid-implementation, calling for help only after the budget is bleeding and timelines have slipped. By then, the foundation has already cracked. The four pillars don't materialize on their own. They must be built deliberately, and the time to start building is earlier than most agencies realize.

The decisions that determine project success or failure are often made before an RFP is ever issued, vendors are invited to demonstrate, or a single contract is signed. The vision you define, the requirements you document, the governance structures you establish, they all become the foundation that carries through procurement, implementation, and beyond. Get this preparation right, and the four pillars have solid ground to stand on. Skip it or rush it, and even the best independent consultant will spend years trying to fix what should have been built correctly from the start.

This critical preparation period is what we call Phase Zero. It separates agencies who control their destiny from those who surrender it before the project even begins.

4

Phase Zero Mastery:
DIY or Independent Jumpstart

WHAT YOU'LL LEARN

- Why the majority of project failures trace back to skipped Phase Zero work
- The 15 essential Phase Zero deliverables that form a foundation for success
- When to DIY vs. engage independent consultants
- How to use pre-RFP demos and market scans to understand the landscape.

THE $185 MILLION MEETING THAT SHOULD NEVER HAVE HAPPENED

The session of the Cook County Board in March 2024 was tense, which is never a good sign. Board President Toni Preckwinkle's face told the story before she spoke: another technology disaster, another budget crisis, another front-page embarrassment for Illinois's second-largest county.

"The Tyler Technologies implementation is now projected at $250 million plus over budget," she announced to the packed chambers, reporters scribbling furiously. "Timeline has extended by 5 years. We're investigating how this happened."[25]

I'll tell you exactly how it happened: Cook County skipped Phase Zero.

They went straight from "we need a new system" to "let's pick a vendor." There was no comprehensive current state assessment, no stakeholder alignment, no process documentation, no readiness evaluation. They confused motion with progress, activity with achievement. They thought they were saving time and money by jumping directly to procurement.

Instead, they discovered what Birmingham discovered, what Rochester Schools discovered, and what dozens of government agencies discover every year. The time you save by skipping Phase Zero is measured in months. The time you lose is measured in years. The money you save is measured in thousands. The money you lose is measured in millions.

Sitting in that packed boardroom, watching commissioners grill IT leadership about how a $30 million project became a $250 million crisis, I kept thinking about Monroe County. They were implementing at the same time, facing similar challenges and complexity. However,

---

25   Based on public records and news reports from outlets such as *The Chicago Tribune* and *Injustice Watch* covering the Cook County integrated property tax system project's budget and timeline issues through 2024.

Monroe County invested 18 months in Phase Zero before they even thought about selecting software.[26] Cook County invested none.

THE MYTH OF STARTING FAST

There's a seductive myth in government IT that starting fast also means finishing fast. That every day spent planning is a day lost to implementation. That preparation is just procrastination dressed up in a suit.

This myth has many authors.

Vendor sales teams feed it, nurture it, weaponize it. "Your peers are already implementing," they whisper. "The longer you wait, the further behind you'll fall. Sign now, and we'll figure out the details during implementation." They manufacture urgency through artificial deadlines: sunset announcements for legacy systems, limited-time upgrade offers, discounts that mysteriously expire at the end of the quarter. The pressure to act now is rarely about your timeline—it's about theirs.

System integrators reinforce the mythology. "We've done this hundreds of times," they assure you. "Our accelerators will compress discovery. Our templates will shortcut requirements. Trust our methodology, and we'll discover what you need as we build."

Even the professional associations (thinking of a large financial officers' organization) and industry groups that should know better sometimes contribute to the problem. Their conferences and training

---

26   Project Management Institute, "Pulse of the Profession 2018: Success in Disruptive Times," *pmi.org*, 2018.

programs often teach procurement and implementation methodologies that haven't evolved in decades, and frameworks developed before cloud computing, AI, and the modern ERP landscape. Well-meaning professionals attend these sessions and return to their agencies with playbooks designed for a different era, confident they're following best practices when they're actually following outdated ones.

Then there's the internal pressure. Your new CIO implemented Oracle at their last agency and is certain they can replicate that success here. Your CFO shepherded a Workday rollout in the private sector and sees no reason why government should be any different. "I've done this before," they'll tell you. "I know what works. We don't need eighteen months of planning; we need to execute." What they don't account for is that every organization is different. The politics, legacy systems, and stakeholder landscape are all different, and the institutional knowledge required to navigate them doesn't transfer from one agency to the next.

Even your own teams might believe the myth. "We've been talking about this for years," they'll say. "It's time for action, not more analysis. Let's pick a vendor and get moving."

Add it all together and you have a perfect storm: vendor pressure, outdated frameworks, executive bravado, and organizational impatience, all pushing toward the same conclusion. Start now. Figure it out later. Move fast and fix things as you go.

But here's what they don't tell you: Industry data consistently proves this is a lie. The Project Management Institute's research reveals that inaccurate requirements gathering is a primary cause of project failure in 35% of cases.[27] Every week invested in preparation prevents months of crisis, and every stakeholder engaged before vendor selection is one less adversary during go-live.

Phase Zero isn't delay; it's acceleration. It's the difference between driving cross-country with GPS versus a vague notion of heading west. You can start driving immediately without a map, but you'll spend far more time lost, backtracking, and asking for directions than if you'd spent an hour planning your route.

WHAT PHASE ZERO REALLY IS (AND ISN'T)

Let me be clear about what Phase Zero is not. It's not analysis paralysis. It's not bureaucratic box-checking. It's not an excuse to avoid hard decisions or delay indefinitely while pursuing perfect information. Phase Zero is the disciplined, systematic preparation that transforms ERP implementation from gambling to engineering. It's the foundation that determines whether your transformation becomes a case study in success or a cautionary tale of failure.

Think of it like preparing for major surgery. You wouldn't let a surgeon operate without diagnostics, blood work, medical history, and surgical planning. You wouldn't accept "we'll figure it out once we open you up" as a strategy. Yet that's exactly what organizations do when they skip Phase Zero. They let vendors perform major organizational surgery without diagnosis, without preparation, without a plan.

---

27   Project Management Institute, "Pulse of the Profession 2021," *pmi.org*, 2021.

A comprehensive Phase Zero accomplishes five critical objectives that can't be achieved any other way.

1. It reveals your true current state. Not the documented processes gathering dust in your procedure manuals, but how work actually gets done, including all the workarounds, exceptions, and undocumented tribal knowledge that keep your organization functioning.
2. It aligns stakeholders before vendor selection. This means getting everyone to agree on problems, priorities, and success metrics before vendor sales teams start making promises and creating competing visions of the future.
3. It identifies your non-negotiables. These are the requirements that absolutely must be met, the processes that cannot change, the constraints that will ultimately determine success or failure regardless of what any vendor promises.
4. It surfaces hidden complexities. Every organization has them: the integrations nobody remembered to mention, reports required by statute, union rules that govern workflows, political dynamics that could derail everything if not addressed early.
5. It builds organizational readiness. This means preparing your people for the magnitude of change ahead, identifying skill gaps before they become crises, and addressing resistance before it metastasizes into outright opposition.

THE 15 ESSENTIAL PHASE ZERO DELIVERABLES

After guiding dozens of government organizations through successful transformations, I've identified 15 deliverables that separate prepared organizations from those stumbling toward disaster. Miss any of these, and you're building on sand.

1: Current State Process Documentation

Not the official processes in your procedures manual. The real processes. How does accounts payable actually work when vendors call angry about late payments? What really happens during year-end close when three people are on vacation and the audit deadline is tomorrow? In one city I worked with, the team insisted their procurement process was fully documented. When we actually observed the process, we discovered 17 undocumented workarounds that had evolved over years to handle edge cases. Hadn't we discovered these during Phase Zero, the new system would have failed on day 1.

2: Future State Vision & Requirements Foundation

Current state documentation tells you where you are. Future state vision defines where you're going. This isn't a wish list of features or a vendor's glossy demo. It's a clear articulation of how your organization will operate after transformation, grounded in strategic priorities and operational realities. What processes will change? What inefficiencies will be eliminated? What new capabilities will you gain? Without this foundation, you'll evaluate vendors against vague notions rather than concrete requirements, and you'll discover mid-implementation that different stakeholders had entirely different destinations in mind.

3: Stakeholder Power Map & Executive Alignment

Every government organization has official organizational charts and unofficial power structures. Phase Zero must map both. Who can kill this project with a phone call? Which department head's resistance could trigger a revolt? But mapping isn't enough. You need explicit, documented commitment from executive sponsors who understand what they're signing up for and have agreed to stay engaged for the duration. Consider a state agency that implemented a new financial system without this groundwork. Six months in, an influential state senator started asking pointed questions because constituents had been complaining, and nobody had thought to brief him proactively. The project survived, but barely.

4: Integration Inventory

Every connection between your current systems and other systems must be documented. This includes the obvious integrations such as banks and state reporting systems, but also the non-obvious ones like the custom feed to the mayor's dashboard or the overnight batch process that nobody quite remembers building. Monroe County discovered 127 integrations during their Phase Zero. They had estimated 40.[28] Each unknown integration discovered during implementation

---

28   Monroe County, NY, "ERP Project Phase Zero Findings Report," Public Document, 2022.

would have been a crisis. Discovered during Phase Zero, they were simply items to plan for.

5: Compliance Requirements Matrix

It includes every law, regulation, rule, and mandate that governs your operations. Birmingham's disaster was partially caused by missing UK statutory reporting requirements during preparation.[29] The system literally couldn't produce legally required reports. This wasn't discovered until after go-live, when it was too late. Your requirements matrix must include federal requirements, state mandates, local ordinances, grant conditions, and union agreements. If it governs how you operate, it belongs in this document.

6: Data Quality Assessment

The brutal truth about your data. Dirty data in a new system is still dirty data, just harder to fix. A county I evaluated had been operating under the comfortable fiction that their data was "pretty clean." Phase Zero revealed that 30% of vendor records were duplicates and 45% of employee addresses were outdated. Discovering this during implementation would have delayed go-live by months. Discovering it during Phase Zero gave them time to clean it before migration even began.

---

29 Based on findings from the UK National Audit Office reports and UK media coverage regarding the Birmingham City Council Oracle implementation failures.

7: Governance Structure

Who makes decisions, and how? This question seems simple until you're mid-implementation and three different committees claim authority over a critical design choice. Phase Zero must establish the steering committee composition, escalation path for disputes, threshold for decisions that require executive involvement, and cadence of governance meetings. It must also clarify the relationship between your internal governance and the vendor's project management structure. Without this clarity, decisions stall, conflicts fester, and accountability dissolves.

8: Change Management & Organizational Readiness

This deliverable answers three critical questions. First, how ready is your organization for change? One city discovered during Phase Zero that they were simultaneously undergoing a school district merger, a major staff reduction, and implementation of a new union contract. Any one of these would stress change capacity; all three together guaranteed failure. They wisely delayed the ERP implementation by a year. Second, what's your communication strategy? How will you keep stakeholders informed, address rumors, celebrate milestones, and maintain momentum across a multiyear journey? Third, what's your overall change management approach? Training plans, resistance mitigation, adoption metrics. All of this must be designed before implementation begins, not improvised along the way.

9: Organizational Design for the Future State

ERP implementation isn't just about technology. It's about organizational transformation. Phase Zero must design the organization you're becoming, not just the system you're implementing. A forward-thinking county used Phase Zero to identify that their future state would require data analysts instead of data entry clerks. They began retraining 18 months before go-live, transforming potential layoffs into career advancement opportunities. Job roles, reporting structures, skill requirements, and team compositions all need to be envisioned before you can build toward them.

10: Business Case with Benefits Realization Plan

These are not vague promises of "transformation" but specific, measurable benefits with owners and timelines. Too many organizations build business cases to justify procurement and then never look at them again. Phase Zero creates a living benefits realization plan that becomes the North Star for the entire initiative. Each benefit needs an owner accountable for achieving it, a baseline measurement, a target, and a timeline. Without this discipline, you'll spend millions and be unable to prove what you gained.

11: Risk Register with Mitigation Strategies

It includes every risk that could derail your implementation, documented with probability, impact, and specific mitigation strategies. Cook County's risk register, if they'd had one, should have identified their history of IT project overruns as a critical risk requiring specific mitigation.[30] Instead, they repeated history. Risks aren't just technical. They're political, organizational, financial, and operational. Phase Zero surfaces them all while you still have time to address them.

12: Budget Reality Check

It is the true cost of implementation, not the vendor's optimistic estimate. This includes the hidden costs that never appear in proposals: backfill for employees pulled onto the project, data cleaning and migration, training development and delivery, postimplementation support, and the inevitable contingency for unknowns. Industry benchmarks consistently show that real project costs are 40%–60% higher than initial vendor estimates due to these overlooked factors.[31] It is better to know the real number during Phase Zero than to discover it through budget amendments and emergency appropriations.

13: Success Metrics Dashboard

How will you know if you're succeeding? You need hard metrics updated regularly: process cycle times, error rates, days to close the books, requisition to payment duration. Whatever matters to your organization, defined and baselined before implementation begins. Without Phase Zero baselines, you can't prove improvement. Without defined success metrics, you can't distinguish genuine progress from vendor spin.

---

30 *Injustice Watch* and *The Chicago Tribune* have extensively reported on the history of Cook County's IT projects, providing a basis for this risk assessment.

31 Based on widely cited industry benchmarks from firms such as Gartner and Panorama Consulting Solutions in their annual ERP reports.

14: The Go/No-Go Decision Framework

Perhaps most importantly, Phase Zero must produce a clear-eyed assessment of whether you should proceed at all. This is the deliverable that gives you something precious: the option to stop before you've committed millions and bet careers. Not every organization is ready. Not every business case holds up under scrutiny. Not every timeline is realistic. Phase Zero should produce an honest recommendation, and leadership should have the courage to act on it, even if that means delay.

15: AI Governance & Readiness Assessment

This deliverable has become essential in the past 2 years, and its absence explains why so many AI-enabled ERP initiatives disappoint. The assessment answers four critical questions that determine whether your organization can responsibly deploy AI capabilities.

First, what is your data governance maturity? AI systems are only as good as the data they consume. Organizations lacking clear data ownership, quality standards, and privacy protocols will find that AI amplifies their data problems rather than solving them. The NIST AI Risk Management Framework calls this the "Govern" function: establishing policies and accountability before any AI system touches your operations.

Second, what AI use cases align with your mission and risk tolerance? Not every process benefits from AI, and some AI applications carry risks that government agencies cannot afford. Detroit's facial recognition disaster, which resulted in a wrongful arrest and $300,000 settlement, happened because the city deployed AI in a high-stakes context without adequate safeguards.[32] Austin canceled $2 million in surveillance technology after public backlash revealed that governance hadn't kept pace with adoption.[33] Phase Zero must map potential AI applications against a risk framework that accounts for civil rights implications, transparency requirements, and community trust.

Third, do you have the organizational capacity for human oversight? The Blueprint for an AI Bill of Rights and emerging federal guidance like OMB M-24-10 require meaningful human review for consequential decisions. This isn't just a checkbox. It requires trained staff, clear escalation protocols, and defined accountability for AI-assisted outcomes. Organizations rushing to implement AI without building this capacity create what researchers call the "Innovation-Governance Gap," where technology adoption outpaces the risk management and policy oversight necessary to use it responsibly.

Fourth, what vendor AI commitments will you require? Modern ERP platforms increasingly embed AI capabilities, but *embedded* doesn't mean *governed*. Phase Zero must define procurement requirements for AI transparency, bias testing, data use restrictions, and audit rights. Vendors should be prohibited from training models on your nonpublic data without explicit consent. You should contractually require explainability for any AI-driven recommendations affecting citizens or employees.

The deliverable itself should include an AI use case inventory mapping potential applications to risk categories, a data readiness scorecard assessing quality and governance across relevant datasets, a

---

32   Office of the Cook County Board President, Board Meeting Minutes, March 14, 2024.

33   Avèro Advisors, "Port of Portland, OR Case Study," *averoadvisors.com*, Accessed September 2025.

human oversight protocol defining who reviews AI outputs and how, and draft procurement clauses addressing AI-specific requirements. Organizations completing this assessment position themselves to benefit from AI acceleration, while avoiding the liability, backlash, and failed investments that plague agencies racing ahead without governance foundations.

THE AI-ACCELERATED PHASE ZERO REVOLUTION

Here's where things get exciting. The Phase Zero process described previously represents best practice refined over the past decade. But the landscape is shifting rapidly, and agencies beginning their Phase Zero work in 2026 face a fundamentally different toolkit than those who started even 2 years ago.

A new generation of AI tools has collapsed the traditional discovery timeline. Where requirements gathering once consumed 8–12 weeks of intensive workshops, interview transcription, and manual documentation, AI-enabled approaches now produce first drafts in ten days or less. Boston Consulting Group estimates that generative AI cuts overall ERP implementation effort by 20%–40%, with nearly all of that savings concentrated in the early discovery and design phases. Total analyst hours per project are dropping by 30%–60%, while requirements completeness is actually improving, with AI-assisted teams identifying 34% more gaps than traditional methods.[34]

---

34   Analyst hours dropping 30%-60%; 34% more gaps identified |Composite: Copilot4DevOps (https://copilot4devops.com/5-ai-tools-for-requirements-management/) for analyst hours; Eltegra (https://www.eltegra.ai/blog/15-ai-prompts-requirements-gathering-business-analysis) for 34% gaps; ArgonDigital (https://argondigital.com/blog/general/how-we-use-ai-to-write-requirements/)

This isn't theoretical. Tools such as Caliber (an innovation from Avero), ClickUp Brain, ScopeMaster, Copilot4DevOps, and Visure AI have been transforming how independent consultants approach discovery. AI engines transcribe every stakeholder workshop, automatically summarizing decisions and action items, while reducing the note-taking errors that plague traditional approaches. Natural language processing models scan draft requirements continuously, identifying duplicates that inflate scope, contradictions that will surface as costly change orders, and missing elements that represent hidden project risk. Requirements flow directly into project management platforms such as Jira or Azure DevOps, creating end-to-end traceability that audit-conscious public agencies desperately need.

However, speed without discipline is dangerous, and this is where the role of independent consultants becomes even more critical in an AI-enabled world.

THE INNOVATION-GOVERNANCE GAP

According to Stanford's 2025 AI Index, AI-related incidents rose to 233 in 2024, a 56.4% year-over-year increase.[35] This surge illustrates what happens when adoption outpaces oversight. Municipal governments are especially vulnerable because residents cannot opt out of essential city services. When a private company deploys flawed AI, customers can switch providers. When a city deploys flawed AI in

---

35   Stanford 2025 AI Index — 233 incidents, 56.4% YOY increase|Stanford HAI. AI Index 2025: State of AI in 10 Charts. https://hai.stanford.edu/ai-index/2025-ai-index-report

permitting, code enforcement, or benefits administration, citizens have no alternative. Mistakes become widespread and unavoidable.

This dynamic creates what I call the Innovation-Governance Gap: the widening chasm between technology adoption, which accelerates exponentially, and risk management and policy oversight, which mature far more slowly. Organizations focused primarily on technology adoption without equally strengthening governance controls and policy guardrails find that real-world harms scale as fast as the technology itself.

Closing this gap requires treating AI governance not as a bureaucratic hurdle but as a strategic investment in public trust and operational resilience. The NIST AI Risk Management Framework provides the essential engineering structure: "Govern" establishes responsible AI policies, "Map" identifies AI risks in context, "Measure" quantifies and assesses those risks through testing and validation, and "Manage" provides continuous monitoring and human-in-the-loop controls. Rights-based frameworks such as the Blueprint for an AI Bill of Rights add the legal guardrails: nondiscrimination protections, data privacy requirements, explanation rights, and human alternatives for consequential decisions.

Phase Zero must embed both pillars. Technical rigor without rights-based boundaries produces efficient systems that harm citizens. Rights-based principles without technical governance produce policies that sound good but lack enforcement mechanisms. The combination produces AI deployment that actually works for the public sector.

THE COMING WAR FOR THE ENTERPRISE BRAIN

Beneath the surface of AI-enabled Phase Zero is a strategic battle that most government agencies don't yet recognize. The major ERP vendors are racing to embed AI capabilities directly into their platforms, positioning themselves as the enterprise brain that will orchestrate not just your financial and HR transactions, but your organization's intelligence layer as well. Oracle, SAP, Workday, and Microsoft are all building AI copilots, intelligent automation, and predictive analytics into their core offerings. The implicit promise is compelling: Implement our platform, and you'll get AI capabilities included.

The alternative model is composable: Rather than accepting a single vendor's vision of your AI future, agencies can layer specialized AI tools onto existing systems, mixing best-of-breed solutions that address specific needs without wholesale platform commitment. For example, a utility district could overlay a rate-modeling AI service onto its legacy billing platform. Gartner calls this "composable ERP," a mosaic of interoperable services that can be swapped or upgraded without disturbing the core ledger.[36]

Why does this matter for Phase Zero? Because the decisions you make during preparation will determine which side of this divide you land on for the next decade. If your requirements are captured using a vendor's proprietary AI tools, shaped by their training data and optimized for their platform, you've already begun the lock-in process

---

36 Gartner "composable ERP" |Public Sector Network (https://publicsectornetwork. com/insight/modernize-government-erp-systems-to-improve-innovation-and-agility). Note: no direct Gartner URL cited in drafts

before you've issued your first RFP. Vendor-led discovery teams often use AI to pre-align requirements to their roadmaps, reviving the risk of covert lock-in under the guise of efficiency.

NASPO and NASCIO's 2024 joint report highlights AI's potential to streamline procurement processes where AI assistants are paired with independent facilitation. That pairing is the key phrase. Independent consultants counter vendor-led AI bias by running parallel discovery processes, cross-checking AI outputs against statutory mandates and business objectives, and documenting the lineage of every requirement so that stakeholders can trace each one back to its origin. This transparent chain follows what the World Economic Forum's AI Procurement Guidelines describe as focusing on a clear problem statement rather than specifying a solution—what we can think of as a 'problemfirst' approach to procurement,"[37] and it's becoming essential as AI reshapes the discovery landscape.

THE FOUR-STEP AI-ACCELERATED MODEL

Leading independent consultants (including Avèro) are weaving AI tools into a repeatable framework that preserves rigor while slashing cycle time. The approach includes the following four components.

1. Rapid Intake with AI Transcription

Every workshop is recorded (where permitted), and AI engines produce structured summaries within hours rather than days. Deci-

---

37   https://www3.weforum.org/docs/WEF_Guidelines_for_AI_Procurement.pdf

sions, action items, and stakeholder concerns are captured with accuracy that manual notetaking cannot match. This alone eliminates the lost-in-translation problem that plagues traditional requirements gathering, where critical nuances disappear between the meeting room and the documentation.

2. Automated Quality and Gap Analyses

Natural Language Processing (NLP) enabled models such as Savoi scan requirements continuously, identifying duplicates that inflate scope, contradictions that will surface as costly change orders, and missing elements that represent hidden project risk. Traditional teams might catch these issues weeks into the process. AI surfaces them immediately, while there's still time to address them without derailing timelines.

3. Human Validation and Stakeholder Alignment

AI produces drafts; humans validate them. This step is nonnegotiable. Independent consultants run structured "red team" sessions where functional leads challenge AI outputs, ensuring that the efficiency of automation doesn't come at the cost of accuracy. Public sector vocabularies such as fund accounting, bond covenants, and grant management rarely appear in commercial AI training data, making manual review essential for government implementations. The EU AI Act now requires "meaningful human oversight" for high-risk systems, and best-practice frameworks stress that AI should augment human judgment rather than replace it.[38]

4. *Living Digital Traceability*

Requirements flow directly into project management and development platforms, creating links to test cases, change orders, and acceptance criteria that persist throughout the project lifecycle. For audit-heavy public agencies, this traceability isn't a nice-to-have. It's essential for demonstrating that the delivered system matches the documented requirements, and it becomes invaluable when questions arise years later about why certain decisions were made.

QUICK WINS VERSUS VENDOR TIMELINES

One of the most significant shifts in the AI era is the emergence of quick-iteration tools that solve problems today rather than waiting years for vendor customizations. Traditional ERP implementations follow a familiar pattern: Identify a gap between your requirements and the vendor's standard functionality, negotiate a customization, wait months or years for development, pay substantial change order fees, and hope the result actually addresses your original need.

---

38   EU AI Act "meaningful human oversight" |IAPP (https://iapp.org/news/a/eu-ai-act-shines-light-on-human-oversight-needs); Dialzara (https://dialzara.com/blog/human-oversight-in-ai-best-practices/)

AI-enabled micro-SaaS solutions offer an alternative. Targeted tools can be deployed in weeks, addressing specific pain points while your core systems remain stable. A state agency used an AI document classifier to automate invoice routing, eliminating a manual bottleneck without touching their legacy financial system. According to GovTech, cities like Tempe are connecting multiple narrow AI agents through orchestration layers that insert mandatory human checkpoints and preserve auditable trails.[39]

Phase Zero must now account for this third option. Your current state assessment should identify not just what needs to be replaced, but what might be extended through targeted AI solutions. Your future state vision should distinguish between capabilities that require core system replacement and those that could be addressed through composable additions. Your vendor evaluation criteria should include assessment of how each platform supports or restricts integration with third-party AI tools. Finally, your AI Governance & Readiness Assessment should evaluate whether your organization has the data quality, oversight capacity, and governance maturity to deploy these tools responsibly.

KEEPING INDEPENDENCE AT THE CENTER

AI's speed is intoxicating, but it magnifies risk when left unchecked. The same tools that accelerate discovery can also accelerate bias,

---

39   Tempe agentic coordination hubs / orchestration layers |GovTech Insider (https://insider.govtech.com/california/sponsored/getting-ai-ready-in-state-and-local-government)

lock-in, and vendor dependency. Over 80% of AI projects stall or fail, twice the rate of ordinary IT initiatives, according to RAND analysts.[40] This failure rate reflects what happens when organizations adopt technology without building the governance foundations necessary to use it effectively.

Independent consultants serve as the essential counterweight to AI's efficiency pressures. They demand that vendors expose how their AI tools make recommendations. They verify that public sector terminology and constraints are properly represented in training data. They ensure that the time savings from AI don't come at the cost of thoroughness and that accelerated timelines don't translate to accelerated disasters.

The agencies that succeed in the AI era will be those that combine both capabilities: using AI to compress timelines while using independent oversight to ensure that speed serves their interests rather than the vendor's. AI without independence is just a faster lock-in. Independence without AI is leaving efficiency on the table. *Phase Zero must position you for both.*

THE PRE-RFP DEMO

During Phase Zero, schedule "educational demonstrations" with major vendors. These aren't sales pitches but are invaluable learning

---

40   80%+ AI projects stall or fail (RAND) |RAND Corporation (https://www.rand.org/content/dam/rand/pubs/research_reports/RRA2600/RRA2680-1/RAND_RRA2680-1.pdf); also Cybersecurity Dive/S&P Global (https://www.cybersecuritydive.com/news/AI-project-fail-data-SPGlobal/742768/)

sessions that reshape how your organization thinks about modernization.

I recently guided a mid-sized city through this process. We scheduled sessions with five vendors over 3 weeks, and what happened surprised everyone, including me.

The accounts payable supervisor who'd spent years insisting their three-way matching process was "completely unique" watched a vendor configure it in 12 minutes using standard functionality. Her skepticism dissolved into curiosity. Staff members who'd been quietly resistant, because they feared learning an entirely new system, became engaged advocates once they could envision the future. Abstract anxiety gave way to concrete possibility.

But the demos revealed something beyond software capabilities—vendor culture. Which sales engineers listened before launching into their scripts? Which vendors brought consultants with actual government implementation experience versus generalists who kept saying "that's similar to what we did for a retail client"? Which ones answered "I don't know, but I'll find out" versus improvising answers that didn't quite fit? These signals matter enormously, and you simply cannot extract them from an RFP response.

Here is a critical caveat: Coordinate closely with your procurement office before scheduling these sessions. Structure them explicitly as educational and market research activities, not as product evaluations, to preserve the integrity of your eventual formal procurement. Document the educational purpose, invite broadly across departments, and avoid any discussions of pricing or specific proposals. Done correctly, these sessions strengthen your procurement; done carelessly, they create protest vulnerabilities that can delay your project by months.

THE DIY VERSUS INDEPENDENT CONSULTANT DECISION

One of the most consequential decisions you'll make is whether to conduct Phase Zero internally or engage independent consultants. I've watched organizations agonize over this choice for weeks, and I've watched others make it reflexively in either direction, sometimes getting it right, but often getting it wrong.

The honest answer is that DIY Phase Zero can absolutely work, but only under specific conditions. You need staff with recent ERP implementation experience, not people who survived an implementation 15 years ago, but those who've guided one in the current technology landscape. You need bandwidth, meaning key personnel who can dedicate substantial time without their regular responsibilities suffering or creating resentment among colleagues picking up the slack. You need political capital to burn, because Phase Zero will surface uncomfortable truths and force difficult conversations; if your IT director is already embattled or your finance team is recovering from a recent controversy, this probably isn't the moment to add fuel. You need timeline flexibility, because learning curves are real and mistakes will happen. And you need manageable complexity—a single-department implementation is different from an enterprise-wide transformation touching every function.

Be brutally honest about whether you meet these criteria. The organizations that successfully execute DIY Phase Zero share a common trait: They assess their readiness with clear eyes rather than hopeful assumptions.

Independent consultants become essential when any of those conditions are absent, but especially when several converge. If your last major system implementation was a decade ago and half the people involved have since retired, you're starting from scratch on institutional knowledge. If your organization is politically complex, with competing fiefdoms or elected officials who view technology projects as campaign ammunition, you need someone who can facilitate without being caught in the crossfire. If your timeline is compressed because a vendor is sunsetting your current system or federal compliance deadlines are looming, you cannot afford the luxury of learning as you go. If key stakeholders are skeptical, perhaps burned by a previous failed project, an outside voice often carries credibility that internal champions simply cannot muster. And paradoxically, if you're resource constrained, trying to layer Phase Zero onto already-stretched staff often produces work so superficial it creates more problems than it solves.

What independent consultants actually provide isn't magic but accumulated pattern recognition, objective facilitation that your internal team cannot provide because they're embedded in your politics, and proven methodologies refined across dozens of implementations. They've seen what works. More importantly, they've seen what fails, and they know how to steer you away from decisions that feel reasonable in the moment but create cascading problems later.

The wrong choice here is expensive either way. Organizations that attempt DIY without the prerequisites often produce Phase Zero deliverables that look complete but lack the rigor to withstand vendor scrutiny or guide real decision-making, which are hollow documents that provide false confidence. Organizations that engage consultants when they genuinely have internal capacity waste money and sometimes breed resentment among capable staff who feel bypassed. Choose deliberately.

THE HIDDEN ROI OF PHASE ZERO

The return on Phase Zero investment is substantial, but it often remains invisible to budget committees and elected officials because it's measured in disasters avoided rather than efficiencies gained. You cannot point to a line item showing "catastrophic failure we prevented." You cannot quantify the change orders that never materialized, the lawsuits that were never filed, the newspaper headlines that never ran. This makes Phase Zero a difficult sell in organizations conditioned to evaluate investments by what they produce rather than what they prevent.

But the evidence, when you examine it honestly, is overwhelming.

Cook County, Illinois skipped comprehensive pre-implementation planning in their enterprise system modernization. The result was a $185 million cost overrun and a 2-year schedule delay that paralyzed county operations and generated withering public scrutiny.[41] What they saved by accelerating past the planning phase, they repaid many times over in emergency fixes, consultant interventions, and operational chaos. The county became a cautionary tale cited in procurement circles nationwide.

The Port of Portland took a different path. They invested in comprehensive Phase Zero planning before issuing their first RFP. That investment has already delivered complete stakeholder alignment across historically siloed departments, a procurement strategy calibrated to their actual needs rather than vendor marketing, and a clear

---

41   $185M cost overrun, 2-year delay (Cook County) | Composite from Sept draft footnotes [8–11]: local news coverage of Cook County/Tyler Technologies. No single clean citation — narrative summary across multiple sources

implementation roadmap with realistic milestones. They haven't yet completed their implementation, but they've positioned themselves to avoid the very disasters that consumed Cook County.[42]

The contrast could not be starker. Cook County's apparent savings from bypassing Phase Zero metastasized into a multimillion-dollar crisis that dominated local news coverage and derailed careers. The Port of Portland's investment functions as insurance—money and time spent now to prevent exponentially larger losses later.

When finance directors push back on Phase Zero costs, I ask them a simple question: Would you rather explain a $200,000 planning investment, or a $185 million recovery effort? The math speaks for itself.

THE PHASE ZERO TIMELINE REALITY

How long should Phase Zero take? The honest answer depends on your organization's starting point, but the traditional reality for most government agencies has been 9–18 months. Complex enterprises with multiple departments, union considerations, and legacy system entanglements tend toward the longer end. Smaller agencies with cleaner organizational structures and stronger internal alignment can move faster.

But as discussed throughout this chapter, organizations that have their foundational elements in place can now complete a comprehen-

---

42 Monroe County positioned to avoid disasters |Avèro Advisors (2023). Monroe County ERP Case Study. https://averoadvisors.com/case-studies/monroe-county

sive Phase Zero in 4–8 months using an AI-accelerated approach. The time savings come not from cutting corners but from eliminating the manual bottlenecks that historically consumed weeks: transcribing stakeholder interviews, cross-referencing requirements for gaps and contradictions, drafting and redrafting RFP language, and building comparison matrices across vendor responses. When AI handles the mechanical work, your team can focus on the judgment calls that actually determine success.

I understand the instinct to view even 4 months as too long. Elected officials want visible progress. Budget cycles create artificial urgency. Peer agencies seem to be moving faster. And vendors will enthusiastically assure you that formal Phase Zero planning is unnecessary, that their implementation methodology includes "discovery" and "requirements gathering" as part of their standard engagement. They'll offer to roll it all into one seamless package, saving you the trouble and expense of independent preparation.

This is precisely the sales pitch that Birmingham and Cook County accepted.

Consider where it led them. Birmingham City Council has spent 7 years on their Oracle implementation and still hasn't achieved stable operations. What began as a 19-million-pound project has consumed 216.5 million pounds and counting, with no clear end in sight. Cook County is now approaching a decade since they began their property tax system modernization, with contracts exceeding $185 million and a trail of delayed tax bills, frozen refunds, and public recrimination. These organizations didn't fail because they planned too much. They failed because they let vendors define the planning process, conflating software installation with organizational transformation and discovery with genuine readiness assessment.

A few months of rigorous, independent planning is not a delay. It is the best investment you can possibly make. The organizations that understand this finish first. The organizations that skip it, or delegate it to parties with financial interests in accelerating implementation, are still struggling years later, explaining to auditors and journalists why their "faster" approach produced such catastrophic results.

▎KEY TAKEAWAYS

- **Phase Zero isn't optional.** It is the foundation that determines whether your implementation becomes a success story or a cautionary tale. Organizations that skip it, or delegate it to vendors eager to accelerate toward billable implementation work, consistently produce the disasters that dominate procurement conferences and audit reports. Those that invest in genuine preparation finish on time, on budget, and with the stakeholder support that sustains value long after go-live.

- **AI can dramatically accelerate Phase Zero, but only with the right prerequisites.** Organizations with clean data foundations, engaged leadership, and realistic expectations can now complete comprehensive preparation in 4–8 months rather than 12–18. However, AI amplifies whatever you feed it, so organizations burdened by political dysfunction or unresolved conflicts will simply arrive at failure faster. The acceleration is real, but the fundamentals still matter.

- **Vendor demonstrations during Phase Zero aren't sales pitches.** They are education sessions that reshape how your organization thinks about modernization, surface assumptions that would otherwise remain hidden until implementation and reveal vendor culture in ways that RFP responses never can. Schedule them early, invite your skeptics, and pay close attention to which vendors listen before they lecture.

▌YOUR IMMEDIATE ACTION

Within the next 2 weeks, initiate your Phase Zero planning by assessing your AI readiness using the framework in this chapter, scheduling educational demonstrations with at least three major vendors, identifying your 15 deliverables with preliminary ownership assigned to each, and convening your first stakeholder alignment conversation. Don't wait for perfect conditions or the next budget cycle, because every month of delay allows your legacy systems to accumulate more technical debt, while institutional knowledge continues walking out the door through retirements. The best time to start Phase Zero was a year ago, and the second best time is now.

5

Accelerated Requirements: AI Enabled and Human-Centric

WHAT YOU'LL LEARN

In the previous chapter, we explored Phase Zero as the foundation for implementation success, including how AI acceleration can compress timelines from 12–18 down to 4–8 months. Requirements gathering is where that acceleration becomes most tangible. This chapter takes you deeper into the mechanics:

- How AI collapses requirements gathering from months to weeks
- The real-world tools making this possible
- The critical importance of balancing AI speed with human validation
- How to spot and fill requirement gaps before they become multimillion dollar change orders

THE REQUIREMENTS MEETING FROM HELL

Picture a windowless conference room in a state office building. It's Month 4 of requirements gathering. The walls are covered with sticky notes that lost their stick weeks ago, now held up by tape and determination. The same 12 people who've been meeting every Tuesday for 4 months sit around a table covered in binders and the remnants of working lunches.

"So, let's go through the purchase order approval workflow . . . again," says the facilitator, a consultant billing $300 an hour to document what everyone in the room already knows.

The purchasing director sighs audibly. "We've documented this three times already. It changes based on amount, funding source, commodity code, vendor status, and about 17 other variables I can't remember right now because my brain is melting."

This scene played out in government projects everywhere just a few years ago. The traditional approach to requirements gathering involved endless meetings, manual documentation, and consultants translating business speak into technical requirements through a painful game of telephone stretched across months. It was common for such efforts to consume 9 months and millions in consulting fees and staff time, only to produce a requirements document that was obsolete before the ink dried as new legislation or policy changes invalidated portions of the work.[43]

---

43 Multiple government audit reports and industry benchmarks from Standish Group's CHAOS reports consistently show that lengthy, traditional requirements gathering phases correlate with higher project-failure rates.

Now picture a different scene in the same conference room, but during Week 2 of requirements gathering. Only six people are present, but they're the right six—the ones who actually know how the work gets done. Instead of sticky notes, a large screen displays real-time requirement generation. Instead of a facilitator asking the same questions repeatedly, an AI assistant learns from every answer and flags inconsistencies as they emerge.

This is the modern reality. Organizations that embrace AI-enabled requirements gathering are achieving in weeks what used to take nearly a year, producing validated, consistent, traceable requirements that reflect how work gets done, at a fraction of the cost.[44] This isn't a future possibility; it's happening now in agencies that chose consultants that have moved beyond traditional approaches.

THE REVOLUTION HIDING IN PLAIN SIGHT

The transformation in requirements gathering isn't just about speed. It's about fundamentally reimagining how we capture what organizations need versus what they think they want. Traditional requirements gathering is built on a flawed premise: that people can accurately describe complex processes they perform unconsciously. Ask an accounts payable clerk to describe their workflow, and they'll give you the official version, the one in the procedure manual. They won't mention the workarounds they've developed over 15 years, the

---

44   National Association of State Chief Information Officers (NASCIO), "The AI Advantage in State IT Projects," 2024.

informal approvals that happen in hallways, or the exceptions that occur so rarely they've forgotten they exist until implementation day when everything breaks.

AI shatters this assumption. Instead of relying solely on people describing their work, a well-trained AI tool observes patterns across interviews and identifies inconsistencies. Instead of consultants interpreting what they heard through their own assumptions, AI cross-references statements and flags conflicts. Instead of static documentation that becomes outdated the moment it's printed, AI creates living requirements that evolve as understanding deepens.

Consider what happened when a major Western city prepared for a systems modernization. In the traditional model, consultants would have spent months conducting sequential interviews: the procurement team describing their workflows on Monday, finance explaining their controls on Tuesday, and by Friday, the critical interconnections between departments would be lost in separate documents that nobody would cross-reference until implementation revealed the gaps.[45]

This city tried something different. They ran parallel requirements sessions, each recorded and transcribed by AI. The technology didn't just capture words—it understood context, identified speakers by role, and flagged when requirements from one session conflicted with another.

The real value emerged in the analysis phase. The AI processing these transcripts identified something remarkable: seven different

---

45   City of Denver, Department of Technology Services, "Modernization Program: Lessons from Phase Zero," Public Report, 2024.

departments mentioned needing an informal "approval from finance" for certain transactions, but this workflow wasn't documented in any official process. In traditional requirements gathering, this would have been missed entirely, discovered only during implementation when the system couldn't accommodate an approval path that didn't officially exist, triggering expensive change orders and finger-pointing about who failed to document the requirement. The thoughtfully curated AI product caught this discrepancy in Week 2. A single 1-hour meeting resolved what would have become a six-figure implementation crisis.[46]

THE PATTERN RECOGNITION REVOLUTION

Here's where AI transforms requirements gathering from documentation to discovery. Organizations don't always know what they need because they don't fully understand what they do. The institutional knowledge that keeps operations running is distributed across hundreds of employees, embedded in tribal knowledge that's never been written down, and encoded in workarounds that have become so routine nobody recognizes them as deviations from official process. Requirements gathering isn't about documenting the known—it's about discovering the unknown.

A Pacific Northwest state learned this lesson when they used AI to analyze requirements for their health and human services system.

---

46   Deloitte, "AI in the Public Sector: A Revolution in the Making," 2024.

Leadership believed they understood their operations completely, having run the same programs for decades.[47] Then they fed their existing documentation, interview transcripts, and process maps into an AI analysis platform.

The AI's first discovery shocked everyone: Two departments had completely contradictory, yet both legally mandated, requirements for client data privacy. Health Services required certain data to be immediately accessible for emergency treatment decisions. Legal Services required the same data to be locked pending case resolution to prevent evidence tampering. Both requirements were correct. Both were legally mandated. And nobody had ever noticed they were mutually exclusive because the departments had never sat in the same requirements session.

Traditional requirements gathering would have documented both requirements faithfully. The conflict would have been discovered by developers mid-project, triggering months of escalations, legal consultations, and delays while leadership figured out which law to prioritize. The AI caught this conflict in Week 3. More importantly, it articulated the conflict's precise nature in terms both departments could understand. Within days, the departments met and developed an elegant solution involving "break-glass" protocols with comprehensive audit trails for medical emergencies. The resolution took one meeting and cost nothing beyond the time of the participants.

---

47   Oregon Department of Human Services, "Integrated Eligibility System Modernization: A Case Study in Requirements Validation," 2023.

THE GHOST REQUIREMENTS THAT KILL PROJECTS

It's not what you document that kills projects. It's what you miss—the requirements nobody thinks to mention because they're so obvious that stating them feels redundant, so routine that they've become invisible, or so rare that the people who remember them have retired.

This is where AI becomes almost prescient. By analyzing patterns across thousands of government ERP implementations, AI platforms have identified what consistently gets missed. Disaster recovery requirements, for instance, appear in fewer than half of initial requirements documents because everyone assumes they're "standard" when in fact they need explicit definition: What's the acceptable recovery time? Which systems get priority? Who has authority to declare a disaster? Audit trail requirements suffer the same fate, with agencies assuming vendors will provide adequate logging when in fact the specific retention periods, access controls, and reporting formats required by state law need explicit specification.[48]

When a major Midwestern city used an AI platform for their requirements gathering, the system identified over 100 missing requirements that their manual process had overlooked.[49] Each one represented a potential change order during implementation. The Project Management Institute consistently finds that uncontrolled scope creep, often driven by requirements discovered mid-project, ranks among the primary reasons projects fail. By identifying these

---

48 Based on anonymized data from requirements analysis platforms used in public sector projects. Names of specific platforms are withheld, but the pattern of missed requirements (disaster recovery, audit trails, etc.) is a consistent finding.

49 Project Management Institute, "Pulse of the Profession 2023: Powering Success with PMI," 2023.

"ghost requirements" early, the AI enabled tool prevented millions in potential overruns before a single line of code was written.

The real sophistication isn't just identifying generic gaps that apply to every implementation. It's understanding context. The AI recognizes that a city needs different tax collection requirements than a county, that California agencies need different privacy rules than Texas agencies, and that a jurisdiction with strong union contracts needs different workflow flexibility than one without. One CIO described it as "having a consultant who'd worked in our organization for 20 years and remembered every conversation, every exception, every workaround that everyone else had forgotten."

THE PROTOTYPE REVOLUTION

Traditional requirements gathering produces documents that users sign off on without truly understanding. They read descriptions of screens they've never seen, approve workflows they've never walked through, and validate data models they can't visualize. Then, implementation begins, and the gap between what they approved and what they actually needed becomes painfully apparent.

AI enables something revolutionary: rapid prototype generation. Instead of describing a future system interpreted by various stakeholders differently, you can show users a functional mock-up they can actually interact with.

A large California county pioneered this approach during their recent modernization.[50] As requirements were documented, an AI

---

50   San Bernardino County, IT Department, "ERP Modernization: A Prototyping-Driven Approach," 2024.

tool generated interactive prototypes that same week. Users could click through workflows, see how data would flow between screens, and experience what their future system would actually feel like in daily use.

The revelations were immediate and invaluable. A procurement workflow that looked elegant on paper proved to be a productivity disaster in the prototype, requiring 17 clicks for a simple approval that currently took 3. A report format that met every documented requirement was practically worthless when users saw the prototype and realized the data they actually needed was buried on page 4. An approval matrix that made perfect sense in theory created impossible bottlenecks in practice when users traced a typical transaction through the prototype and discovered it would require eight separate approvals from people who were rarely in the office simultaneously.

Discovering these issues during requirements gathering cost nothing to fix beyond the time spent in the room. Discovering them during implementation would have cost millions in change orders and months in delays while developers reworked functionality that had already been built, tested, and signed off.

THE HUMAN ELEMENT: WHY AI ALONE FAILS

Here's what vendors selling AI requirements tools won't tell you: AI without human judgment is dangerous. It can optimize for efficiency while destroying effectiveness. It can streamline processes while eliminating the human discretion that makes government services humane. It can recommend technically perfect solutions that are politically catastrophic.

I witnessed this firsthand in a state agency that was seduced by a vendor's promise of "fully automated requirements generation." The AI analyzed their benefits determination process and recommended comprehensive automation of eligibility decisions. The business case was compelling: technically feasible, legally compliant with all documented regulations, and projected efficiency improvements exceeding 400%.[51]

However, the human cost would have been catastrophic. Complex situations requiring judgment, such as a veteran with PTSD whose documentation was incomplete, a family fleeing domestic violence without time to gather paperwork, or an elderly person with dementia who couldn't navigate the application process, would have been denied by an algorithm optimized for efficiency at the cost of real outcomes. The political fallout from the first news story about a vulnerable person denied benefits by a robot would have been devastating, and the agency would have spent years rebuilding public trust.

This is why the human-centric component of AI-enabled requirements gathering isn't a nice-to-have. It's essential. Every AI-generated requirement needs human validation across three dimensions: operational feasibility (Can we actually do this with our staff and resources?), political acceptability (Will elected officials and the public support this approach?), and human impact (What happens to the edge cases, the exceptions, the vulnerable populations that don't fit neatly into algorithmic categories?).

The organizations getting this right treat AI as a tool that accelerates human intelligence rather than a replacement. They use AI to identify

---

51 *Government Technology* magazine has published numerous articles and case studies on the ethical challenges and pitfalls of AI in benefits determination, reflecting the scenario described.

patterns, surface conflicts, and generate drafts, but they insist on human review before any requirement becomes final. This balance is what separates successful AI-enabled requirements gathering from the cautionary tales that will inevitably emerge as agencies rush to adopt these tools without appropriate safeguards.

THE SPEED DIVIDEND

When you compress requirements gathering from 9 months to 6 weeks, you fundamentally change the project dynamic in ways that extend far beyond the calendar.

In a traditional timeline, the organization you documented 9 months ago no longer exists by the time you finish. Key stakeholders have moved to other positions or retired, regulations have evolved with new legislative sessions, technology has advanced with new vendor releases, and organizational priorities have shifted with new leadership or budget realities. Your requirements document becomes a historical artifact describing an organization that was, not the organization that is.

But when requirements gathering takes 6 weeks, you're documenting the organization that actually exists. The stakeholders who participated are still engaged and remember what they agreed to. The regulations you documented are still current. The technology landscape hasn't fundamentally shifted. Your requirements remain fresh, accurate, and defensible.

A New England state discovered this during their 2024 ERP implementation.[52] They used an AI-enabled approach to compress what would traditionally have been an 8-month requirements effort into 5 weeks. Because the requirements were fresh, vendors could bid on what the agency actually needed rather than what it had needed 8 months earlier. Because stakeholders remained engaged throughout implementation, disputes about "what we really meant" virtually disappeared. Change orders dropped by an estimated 80% compared to the state's previous major implementation.

The speed also enabled iterative refinement that traditional approaches cannot accommodate. When federal regulations changed mid-project, what would have been a 6-month change order process in a traditional implementation became a 2-day requirements update. The compressed timeline meant the requirements team was still available, the context was still fresh, and the adjustment could be incorporated without the archaeological expedition typically required to understand what had been documented months or years earlier.

CONNECTING TO PHASE ZERO

Everything in this chapter depends on the Phase Zero work we discussed previously. AI-enabled requirements gathering doesn't replace Phase Zero preparation—it builds on it. The stakeholder alignment you achieved in Phase Zero determines whether the right people show up to requirements sessions. The AI readiness

---

52   State of Maine, Office of Information Technology, "ERP Implementation: A Case Study in Accelerated Requirements," 2024.

assessment you conducted determines whether you have the data foundation to support AI-enabled analysis. The governance structure you established determines who has authority to resolve the conflicts that AI will inevitably surface.

Organizations that attempt AI-enabled requirements gathering without Phase Zero preparation consistently struggle. They have the tools but not the organizational readiness to use them effectively. They can generate requirements quickly but can't validate them because stakeholders aren't aligned. They can identify conflicts but can't resolve them because governance structures don't exist. The technology works, but the organization fails.

This is another reason why vendors who want to skip Phase Zero and proceed directly to implementation should concern you. They may offer AI-enabled requirements gathering as part of their methodology, but without the independent preparation that ensures your organization is ready to participate meaningfully, you're simply accelerating toward failure.

▌ KEY TAKEAWAYS

- **The revolution is here.** AI-enabled, human-centric requirements gathering collapses timelines from months to weeks, prevents millions in change orders by surfacing conflicts and gaps early, and produces better results than traditional methods ever could. Organizations still using the old approach are not just moving slower but are producing inferior requirements that will haunt them throughout implementation.

- **Missing requirements are the hidden killers of projects.** AI tools can identify these gaps by comparing your needs against patterns from thousands of similar implementations, catching the ghost requirements that traditional methods miss because they rely on humans remembering to mention things they've long since stopped noticing. Every ghost requirement caught during requirements gathering is a change order prevented during implementation.
- **Prototyping requirements prevents million-dollar misunderstandings.** Letting users experience their future system during requirements gathering is the ultimate form of validation, revealing usability problems, workflow bottlenecks, and conceptual misalignments that no amount of documentation review would surface. Fixing a prototype costs hours; fixing a built system costs months and millions.

▎YOUR IMMEDIATE ACTION

Before your next major requirements effort, audit what happened last time. Calculate the true duration, including all the "just one more session" extensions that stretched weeks into months. Tally the actual cost in staff time diverted from regular duties, consultant fees that exceeded original estimates, and change orders during implementation that traced back to missed or misunderstood requirements. That total represents your baseline, the cost of continuing with traditional approaches.

Next, identify AI tools worth piloting. You don't need a massive procurement to begin. Many AI transcription and analysis platforms offer trial periods or pilot programs specifically designed for government evaluation. Pick one business process, use AI transcription in your next requirements session, and let the tool generate a draft requirements document. Compare what the AI produces to what your traditional documentation captured.

The difference will convince you more than any case study or vendor presentation could. The transformation in requirements gathering isn't a future possibility you should monitor. It's a present reality you should embrace, because the organizations that master these tools now will complete their implementations, while those still using traditional methods are still stuck in Month 4 of requirements gathering, surrounded by sticky notes that have long since lost their stick.

6

Crafting a Winning
Procurement Strategy
in the AI Era

WHAT YOU'LL LEARN

The Phase Zero deliverables you developed and the AI-enabled requirements you gathered aren't endpoints. They're fuel for procurement success. This chapter shows you how to transform that preparation into a procurement process that attracts qualified vendors, survives protests, and produces contracts that actually protect your interests. You'll learn:

- How AI can transform and accelerate RFP development
- The truth about cooperative contracts and when they make sense
- How to build evaluation criteria that surface genuine capabilities rather than polished marketing

- Why everything you accomplished in Phase Zero determines whether your procurement succeeds or fails

THE PROCUREMENT DIRECTOR'S DILEMMA

The California Department of Technology's 2019 ERP procurement should have been a model for the nation. They had the budget, with $500 million allocated. They had the expertise, with some of the state's best IT minds on the team. They had the time, with 2 years of planning. What they didn't have was independence from vendor influence.[53]

By the time the procurement was canceled in 2020, the state had spent millions on consultants who were supposed to help them select an unbiased solution. The problem was that those consultants had deep financial relationships with the very vendors bidding on the project. The requirements they helped craft were so specific, so precisely aligned with one vendor's capabilities, that competing vendors cried foul before proposals were even submitted.[54]

The state auditor's report was damning: "The procurement process was compromised from the start by advisors who had inherent conflicts of interest."[55]

---

53   California Department of Technology, "FI$Cal Project Reports," Accessed 2024.

54   *The Sacramento Bee*, "California spent $50 million on a state tech project. Now it's canceled," March 11, 2020.

55   California State Auditor, Report 2019-112, "Department of Technology: It Must Provide Better Oversight of State Technology Projects," February 2020.

This scenario plays out with variations across governments every year. Well-intentioned procurement professionals, following established processes, somehow produce RFPs that favor specific vendors, exclude capable alternatives, or create requirements that guarantee expensive customization. It's rarely corruption in the traditional sense. It's the predictable result of a procurement process that hasn't evolved to match modern technology complexity and vendor sophistication. Vendors have entire teams dedicated to shaping requirements in their favor, while government procurement offices are stretched thin just managing compliance. The playing field isn't level, and pretending otherwise guarantees failure.

Meanwhile, in Pennsylvania, the Department of Human Services was taking a completely different approach to their integrated eligibility system procurement. They deployed modern data analysis tools, engaged independent consultants with no vendor ties, and created an evaluation process focused on outcomes rather than features. The result was a successful selection that attracted multiple qualified bidders and survived all protests.[56]

The contrast could not be starker. California's traditional approach led to failure and waste. Pennsylvania's modern, independently overseen approach led to success. The difference wasn't resources or expertise. It was approach, specifically the recognition that procurement independence isn't a luxury but a necessity.

---

56   Pennsylvania Department of Human Services, "Integrated Eligibility System Procurement Overview," Public Document, 2022.

THE AI REVOLUTION IN PROCUREMENT

Here's what California didn't implement but Pennsylvania did: AI has fundamentally transformed how smart governments approach ERP procurement—not the buzzword AI that vendors splash across their marketing materials, but practical AI tools that level the playing field between government buyers and sophisticated vendor sales machines.

The transformation builds directly on the AI-enabled requirements work discussed in the previous chapter. Remember how AI collapsed requirements gathering from months to weeks while improving quality? The same acceleration applies to procurement, but with an important caveat. According to industry analysts at Gartner, traditional government IT procurement cycles take 9–18 months.[57] AI-assisted processes can significantly shorten this timeline while improving outcome quality. But AI accelerates good procurement practices; it cannot fix bad ones. If your Phase Zero work was superficial, if your requirements were incomplete, and if your stakeholder alignment was illusory, then AI will simply help you fail faster.

Consider how the City of Boston approached their technology procurement in 2023.[58] Facing the challenge of replacing multiple aging systems, they didn't jump straight to an RFP. They invested in comprehensive Phase Zero preparation, developed AI-enabled requirements using the approaches discussed previously, and then deployed AI tools strategically throughout procurement. Their procurement team used natural language processing to analyze successful RFPs from peer cities with similar demographics and service profiles.

---

57   Gartner, "Benchmark Analytics: IT Key Metrics Data," 2023.

58   McKinsey & Company, "Generative AI: The new productivity frontier in procurement," May 2023.

Machine learning algorithms compared their requirements against vendor capabilities documented in public contracts, revealing which requirements would drive expensive customization and which could be met with standard functionality.[59]

The result was an RFP that attracted six qualified bidders, survived all protests, and led to a selection proceeding on budget. The entire procurement process took a fraction of the typical timeline.[60] But here's what Boston understood that California missed: AI in procurement isn't about replacing human judgment but about augmenting it with pattern recognition and analysis capabilities that no human team could match. The technology accelerated the process, but humans drove every significant decision.

THE FIVE-STEP AI-ENHANCED PROCUREMENT ROADMAP

After helping many governments modernize their procurement, I've developed a five-step roadmap that uses AI without losing the human judgment that ultimately determines success. Each step builds on the Phase Zero groundwork and the AI-enabled requirements we have already outlined. When I say "AI" or "AI tool," I mean a purpose-built, deeply trained system with strong technical and ethical guardrails— far beyond what offtheshelf tools like ChatGPT or Grok can offer.

---

59   City of Boston, Department of Innovation and Technology, "Annual Modernization Report," 2023.

60   City of Boston, "Successful Technology Procurement Announcement," Public Release, 2023.

For this strategy to work, the AI must function as a true "government procurement superbrain".

Step 1: Maturity Assessment

Before you write a single RFP requirement, you need brutal honesty about your organization's readiness—not the political version where everyone claims to be "ready for transformation," but the real assessment that acknowledges gaps, limitations, and areas where preparation fell short.

This is where your Phase Zero deliverables become invaluable. The AI Governance and Readiness Assessment you developed, the Stakeholder Analysis and Engagement Strategy you documented, the Current State Assessment you completed: These artifacts tell the truth about your organization's actual readiness, not the aspirational version that makes everyone comfortable.

AI tools can deepen this analysis in ways that seem almost magical. Feed your Phase Zero outputs into an AI model and ask it to assess procurement readiness. Upload your process documentation, stakeholder feedback, risk assessments, and requirements documents. The AI will identify patterns humans might miss, such as, that every department claiming to be "ready for change" also requested that their current processes remain completely untouched, or that your requirements contain 17 implicit assumptions about data quality your Current State Assessment suggests are unfounded.

A Midwestern state agency used this approach recently and discovered their standardized procurement process actually had 43 variations depending on funding source, amount, and department.[61] The AI tool didn't just identify the variations; it mapped which ones were legally required versus those that existed simply because "that's how we've always done it." This insight saved them from writing requirements for complexity they didn't need, which reduced their RFP scope by nearly 30% before they issued it.

According to the Standish Group, fewer than a third of government IT projects succeed, which suggests that readiness is consistently overestimated.[62] Your maturity assessment is your opportunity to be honest before the procurement process locks you into commitments your organization cannot fulfill.

Step 2: Designing an AI-Augmented RFP Process

The traditional RFP process is broken in ways that favor vendors over government buyers. It takes too long, creating opportunities for requirements to become stale. It rewards vendors who game the system with polished responses rather than genuine capabilities. It produces documents so generic that vendors can copy-paste their way to apparent compliance.

Start with requirements generation. Instead of copying boilerplate from other RFPs or letting vendor-affiliated consultants shape your needs, use AI tools or consulting teams that have proprietary

---

61   National Association of State Chief Information Officers (NASCIO), "State IT Procurement: A Digital Upgrade," 2024.

62   The Standish Group, "2020 CHAOS Report: With a Special Government Addendum," 2020.

AI enabled tools to analyze your Phase Zero outputs and the requirements you developed in the previous phase. The AI tools can transform those requirements into RFP language that actually matches your needs rather than industry templates that may not fit your context.

But here's where it gets interesting: You can also use AI to analyze how vendors will respond. The State of Texas used AI to simulate vendor responses to their draft requirements for a health and human services system.[63] With remarkable accuracy, the AI predicted that certain requirements would trigger "custom development" responses from most vendors. Those requirements weren't complex—they were just phrased in ways that gave vendors cover to propose expensive customization. The state rewrote those requirements to focus on outcomes rather than prescribed approaches, and the final responses included significantly less customization at significantly lower cost.

The City of Austin took this even further, using AI to analyze 5 years of vendor protests across Texas, identifying the requirement patterns most likely to trigger challenges.[64] They adjusted their RFP accordingly and received zero protests, saving months of delay and uncertainty that protests typically create.

The real revolution is in structuring RFPs to reveal truth rather than test compliance. Instead of asking "Describe your implementation methodology," which invites canned responses, ask "Using our specific Phase Zero process documentation attached as Exhibit A, identify

---

63   Texas Department of Information Resources, "AI in Procurement Pilot Program," Case Study, 2023.

64   City of Austin, "Procurement Innovation: Zero-Protest Initiative," 2024.

the three highest-risk integration points and explain your mitigation approach." This kind of scenario-based question requires vendors to engage with your specific situation. AI can generate hundreds of these targeted questions based on your Phase Zero deliverables and requirements documents. Vendors cannot copy-paste their way through questions that reference your specific organizational context.

Step 3: Evaluating Vendors with Unbiased AI Tools

This is where AI truly shines—cutting through vendor marketing speak to reveal actual commitments. The traditional evaluation process is exhausting and inconsistent. Evaluation teams read hundreds of pages of proposals, trying to compare responses that use different terminology, structure information differently, and bury weaknesses in confident-sounding language. Fatigue sets in, important distinctions get missed, and vendors who write well score better than vendors who perform well.

AI changes this dynamic fundamentally. Natural language processing can analyze proposals for concrete commitments versus vague promises, flagging responses that sound good but actually commit to nothing.

The State of Colorado used AI to analyze vendor proposals for their unemployment insurance system modernization.[65] The AI flagged that one vendor had used the phrase "will work with the state to determine" 47 times, essentially admitting they didn't have solutions for 47 requirements. Human evaluators had marked this proposal

---

65   Colorado Governor>s Office of Information Technology, "AI-Assisted Procurement Analysis," Case Study, 2023.

as "highly responsive" because it addressed every requirement with professional-sounding language. The AI revealed it committed to almost nothing specific. That insight changed the evaluation outcome entirely.

But here's where independent oversight becomes crucial. Vendor partners often provide "evaluation assistance" that subtly steers selections toward their preferred outcomes. The evaluators may not even realize they're being influenced; the guidance seems helpful and professional. A major airport in the US avoided this trap by using independent consultants to manage their AI-assisted evaluation.[66] The AI did the analytical heavy lifting, identifying patterns and inconsistencies across proposals, but independent consultants ensured the evaluation criteria remained objective and vendor influence stayed out of the room where decisions were made.

Remember the vendor demonstrations you conducted during Phase Zero? Those observations become invaluable now. You saw which vendors listened before launching into their scripts. You watched which ones brought government implementation experts versus generalists. Those qualitative insights, combined with AI-driven proposal analysis, give you a complete picture that neither approach could provide alone.

Step 4: Negotiating Contracts with AI-Driven Risk Analysis

Contract negotiation is where good procurements often go bad. Government negotiators, even experienced ones, typically negotiate a

---

66   Avèro Advisors, "Port of Portland, OR Case Study," averoadvisors.com, Accessed 2025.

handful of major contracts per year. Vendor contract teams negotiate dozens, and they've optimized every clause to protect vendor interests while appearing balanced. The information asymmetry is enormous.

AI levels this playing field. Tools can compare vendor-proposed terms against hundreds of similar government contracts, identifying clauses that create unusual risk or deviate from market norms in ways that disadvantage you.

The State of Michigan used AI to analyze their vendor's proposed contract against 50 similar government ERP contracts.[67] The AI identified 17 outlier clauses, including limitation of liability terms that were far more restrictive than typical government contracts and intellectual property provisions that would have made switching vendors nearly impossible. Armed with this data and the specific language from comparable contracts, they negotiated terms that protected state interests.

But the real power of AI in contract negotiation isn't just identifying bad terms but predicting future problems. Research by organizations such as the Brookings Institution has analyzed which contract terms correlate with project success.[68] Contracts that tie payments to outcomes, such as user adoption rates or specific performance metrics, succeed more often than those tied only to technical milestones such as "system testing complete." Contracts with specific remedies for vendor performance failures succeed more often than those with vague "good faith" resolution procedures. AI can identify

---

67   Michigan Department of Technology, Management & Budget, "AI Contract Analysis Pilot Results," 2024.

68   The Brookings Institution, "Driving Public Sector Success Through Outcome-Based Contracting," June 2022.

these patterns across thousands of contracts and help you negotiate terms that predict success.

This idea connects directly to the governance structure you established during Phase Zero. Your Executive Steering Committee charter, your escalation procedures, your decision rights framework all need to be reflected in contract terms that give them teeth. A governance structure without contractual authority is advisory at best.

Step 5: Implementing Ongoing Oversight for Performance

The procurement process doesn't end when the contract is signed. That's when the real work begins, and traditional contract management approaches are dangerously inadequate. Quarterly reviews, the standard government approach, often identify problems only after they've metastasized into crises. By the time a status report acknowledges trouble, the vendor has known about it for months and has been managing the narrative rather than fixing the problem.

AI-enabled oversight changes this dynamic. Well-designed and trained "agents" can monitor project velocity in real time, analyzing status reports, meeting minutes, code commits, and ticket resolution rates for early warning signs that match historical failure patterns.

The City of San Diego implemented an AI-powered governance dashboard for their ERP implementation that continuously monitored dozens of metrics rather than waiting for monthly or quarterly reviews.[69] The AI flagged a 40% drop in code commits, while support ticket volume simultaneously increased—a pattern that historically

---

69   The Urban Institute, "Does Contract Language Predict Project Success? An Analysis of Government IT Projects," 2023.

correlates with senior technical staff being reassigned to other projects. The city investigated immediately and discovered that the vendor had quietly moved three senior developers to a higher-priority commercial client. They addressed it within days, months before the impact would have appeared as a schedule delay in traditional status reporting.

This ongoing oversight connects to the Risk Register and Mitigation Strategies you developed during Phase Zero. Those weren't documents to file and forget but were predictions about what might go wrong. AI-enabled monitoring lets you watch for those predicted risks in real time rather than discovering them after the damage is done.

THE COOPERATIVE CONTRACT QUESTION

Now, let's talk about a tool that too many procurement professionals overlook or dismiss—cooperative purchasing agreements. Organizations such as EdgeMarket, OMNIA Partners, and Sourcewell offer access to pre-competed contracts that can be adopted without running your own procurement.[70] The appeal is significant. You save months of procurement time, avoid protest risk, reduce administrative burden, and benefit from competitive pricing already negotiated through rigorous processes.

The conventional wisdom suggests cooperative contracts work only for commodities, not for complex services. That wisdom deserves reconsideration, particularly when it comes to consulting and advisory engagements.

---

70   San Mateo County, "Risk-Adjusted Contracting Model for IT Procurements," 2024.

Think about what a competitive RFP process accomplishes. You spend 3–6 months developing requirements, issuing the RFP, evaluating proposals, and navigating potential protests. During that entire period, your ERP project sits in limbo, your legacy risks compound, and the organizational momentum you've built dissipates. By the time you've selected your independent advisor, the political window for your project may have narrowed or closed entirely.

Now consider what a cooperative contract offers for the same engagement. The cooperative has already vetted the consultant's qualifications, checked references, validated insurance and bonding, and negotiated rates through competitive processes. You can engage qualified advisors within weeks rather than months. Your Phase Zero work can begin while peer agencies are still drafting RFP requirements.

The distinction lies in understanding where cooperative contracts excel versus where they require additional scrutiny. For consulting and advisory services, cooperative contracts often represent the smarter path. The "customization" concern that legitimately applies to software doesn't apply the same way to professional services. An independent consultant adapts their approach to your specific situation regardless of how you engaged them. The cooperative contract gets them in the door. Therefore, your working relationship shapes the engagement.

For major ERP software and system integrator contracts, more caution is necessary. A recent analysis by the Arizona Auditor General examined cooperative contract usage for large software implementations. It was found that initial time savings were sometimes offset by modification delays, and cost assumptions didn't always hold when implementation patterns differed from the original competition.[71]

---

71   City of San Diego, IT Department, "ERP Governance Dashboard Initiative," 2024.

For these high-value, high-complexity purchases, the hybrid approach makes sense: Use cooperative contracts as baselines and benchmarks, while potentially running competitive processes for your specific needs.

Cooperative contracts remove barriers without introducing meaningful risk. The City of Phoenix recognized this distinction.[72] They used competitive processes for their major software decisions, while employing cooperative contracts to quickly engage independent advisors who could guide that competitive process. The advisors were working within weeks, providing the independent oversight that shaped a successful procurement.

When evaluating cooperative contracts for consulting engagements, it is necessary to focus on the cooperative's vetting process. Did they verify relevant government experience? Did they check references from similar engagements? Did they negotiate rates that reflect market reality? Reputable cooperatives conduct rigorous competitions that often exceed what a single agency could accomplish with limited procurement resources.

The procurement professionals who dismiss cooperative contracts entirely are often the same ones who complain about how long it takes to get anything done. The tool exists. It's legally sound. It's been competitively procured. The question isn't whether cooperative contracts are legitimate but whether you'll use available tools strategically or default to processes that create delay without adding value.

---

72   OMNIA Partners and Sourcewell provide lists of available public sector contracts.

THE INDEPENDENT OVERSIGHT IMPERATIVE

Throughout this entire process, one factor consistently separates success from failure—independent oversight. This means having someone in the room whose only financial interest is your success, someone empowered to challenge assumptions, question vendor claims, and say no when necessary.

This isn't about distrusting your internal team but about recognizing the structural disadvantages government buyers face. Vendors have teams dedicated to winning your business, with compensation tied to sales success. They've analyzed thousands of procurements and optimized their approaches accordingly. Your procurement team, no matter how talented, handles a fraction of that volume and cannot match that pattern recognition.

Monroe County understood this throughout their procurement. Independent consultants provided oversight at every stage, from requirements development through vendor selection and contract negotiation.[73] The cost was less than 2% of the project budget. Studies have consistently shown that independent oversight is a key factor in avoiding the 40%–60% cost overruns typical in projects lacking it.[74] That's not a cost—it's insurance with a proven return.

---

73 Arizona Auditor General, "State of Arizona Cooperative Purchasing Review," Report No. 24-105, July 2024.

74 City of Phoenix, Finance Department, "Hybrid Cooperative Procurement Model Pilot," 2023.

WHEN SPEED MATTERS: THE EMERGENCY PROCUREMENT TRAP

Sometimes disasters strike, and you need a new system immediately. Your legacy system has failed catastrophically, a natural disaster has destroyed infrastructure, or a compliance deadline has arrived with no extensions possible. The pressure to skip competitive procurement and move fast is enormous.

But emergency procurement for enterprise systems is almost always a trap. The State of Louisiana learned this after Hurricane Katrina, when emergency-procured systems created decades of vendor lock-in and integration problems that cost far more than a thoughtful procurement would have.[75] The urgency that felt so pressing in the moment led to constraints that persisted for years.

A better path was demonstrated by the City of Paradise, California.[76] After the Camp Fire destroyed their infrastructure, they faced genuine urgency. But instead of emergency-procuring a permanent system under crisis conditions, they deployed cloud-based temporary solutions to maintain essential services while running an accelerated but still competitive procurement for their long-term platform. The result was a better system at lower cost, without the long-term pain of a decision made under duress.

If you find yourself facing emergency procurement pressure, ask whether the emergency is truly about the permanent system or about maintaining services while you procure properly. Usually, temporary solutions can bridge the gap, while you protect your long-term interests.

---

75   Avèro Advisors, "Monroe County, NY Case Study," averoadvisors.com, Accessed 2025.

76   Panorama Consulting Group, "The ERP Report," (Annual).

▌ KEY TAKEAWAYS

- **AI transforms procurement from an administrative burden to a strategic advantage, but only when paired with human judgment and independent oversight.** The technology can analyze proposals, identify risks, simulate vendor responses, and monitor implementation progress in ways that level the playing field with sophisticated vendor sales teams. But AI without independence simply accelerates whatever biases and conflicts already exist in your process. The organizations succeeding with AI-enabled procurement are those that combine technological capability with structural independence.

- **Cooperative contracts are strategic tools that deserve a place in your procurement toolkit, particularly for consulting and advisory engagements.** The months you spend running an RFP for independent oversight are months your project sits in limbo and your legacy risks compound. Reputable cooperatives have already vetted qualifications, checked references and negotiated competitive rates through rigorous processes. For professional services that adapt to your specific needs regardless of engagement method, cooperative contracts remove barriers without introducing meaningful risk. Reserve your competitive procurement energy for the major software and system-integrator decisions where customization concerns genuinely apply, and use cooperative contracts strategically to get independent advisors working while peer agencies are still drafting RFP requirements.

- **Independent oversight throughout procurement is the difference between buying promises and purchasing success.** The investment is minimal, relative to project budgets, but the return is substantial. Organizations with independent oversight consistently avoid the massive overruns that plague government IT procurement. As multiple studies have demonstrated, it's not about whether you can afford independent oversight but whether you can afford the consequences of proceeding without it.

▌ YOUR IMMEDIATE ACTION

Within the next 2 weeks, audit your current procurement approach. Calculate how long your typical technology procurement actually takes, from initial planning through contract execution. Identify how often you face protests and how much delay they create. Examine who in your current process has no vendor relationships whatsoever, and whether that person is empowered to challenge assumptions and protect your interests.

Then, map your Phase Zero outputs to procurement requirements. Your AI Governance and Readiness Assessment should inform your evaluation criteria for vendor AI capabilities. Your Stakeholder Analysis should shape your communication strategy during procurement. Your Requirements Documentation should form the foundation of your RFP. Every deliverable you created has a role in procurement—identify those connections explicitly.

Finally, evaluate AI tools for your next procurement. Focus on three capabilities: requirements-to-RFP generation that transforms your documented needs into procurement language, response analysis that cuts through vendor marketing to identify actual commitments, and contract risk assessment that compares proposed terms against market benchmarks. Many of these tools offer trial periods or pilot programs designed for government evaluation.

Your Phase Zero work built the foundation. Your AI-enabled requirements captured what your organization needs. Now procurement determines whether that preparation translates into a contract that protects your interests or one that locks you into vendor dependency for a decade. The agencies that succeed treat procurement as the strategic moment it is, deploying AI tools to level the playing field, engaging independent advisors through cooperative contracts to maintain momentum, and building evaluation processes that reveal truth rather than reward polished marketing. The agencies that struggle treat procurement as paperwork to complete. Everything you've built so far depends on getting this phase right.

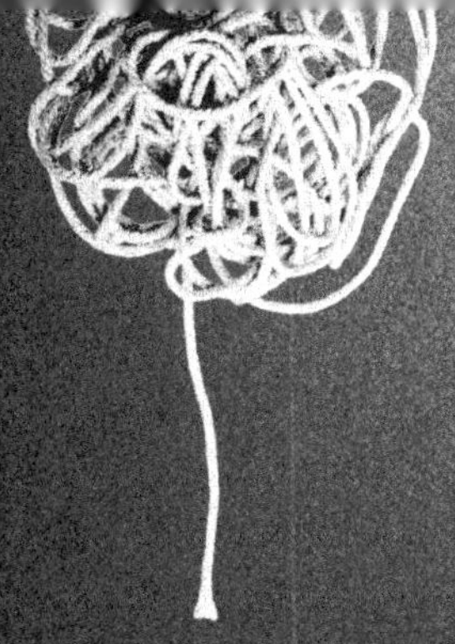

PART III

Executing Without Vendor Capture

Execution is the ultimate test. It is where we structure vendor selection processes to reveal genuine capability rather than rehearsed demonstrations. By aligning vendor incentives with outcomes rather than revenue targets, we create a shared commitment to the results that actually matter.

7

Vendor Selection:
Spotting Allies and Dodging Traps

WHAT YOU'LL LEARN

Your Phase Zero deliverables mapped your organization's reality. Your AI-enabled requirements captured what you actually need rather than what vendors want to sell you, and your procurement strategy created the framework for a fair, defensible selection process. Now comes the moment where those investments either pay off or get squandered—vendor selection.

This chapter illustrates

- How to move beyond generic demonstrations to scenario-based evaluations that reveal genuine capabilities
- The critical difference between vendor-curated references and real-world intelligence you can trust
- The fundamental strategic choice between fully embedded AI systems and flexible composable approaches

- How to negotiate contracts that protect your interests for the decade-long relationship you're about to enter

Organizations that master this phase don't just avoid problematic vendors but identify genuine partners whose success becomes inextricably linked to their own.

THE DEMO THAT LOOKED PERFECT

In 2022, a large Florida county reached the final stages of a $40-million ERP selection, down to two vendors after months of evaluation. Vendor A presented a demonstration that captivated everyone in the room, with every screen beautifully designed, every workflow seamless, and every click producing exactly the expected result. During a break, the finance director turned to her colleagues and said what everyone was thinking: "This is it. This is the future."

Vendor B's demonstration felt fundamentally different. It used standard environments rather than custom-built showcases, occasionally stumbling, and openly acknowledging when specific workflows would require configuration rather than working immediately out of the box. The presentation came across as less impressive but somehow more honest and more grounded in the reality of actual software rather than theatrical performance.

The selection committee was leaning heavily toward Vendor A based on the strength of that flawless demonstration, but the county had engaged independent advisors during Phase Zero, and those advisors urged caution. The lead consultant had witnessed this pattern repeatedly across dozens of government selections and offered an observation that changed the entire conversation: "A perfect demo doesn't mean a perfect system, and in my experience, it often means the opposite. The question isn't whether their demonstration works flawlessly. The question is whether their actual system can handle the complexity of your real business operations."

With their advisors' guidance, the County designed what they called a "Day in the Life" evaluation. It provided both vendors with a sanitized version of actual county data and a series of complex tasks their staff performed daily: processing a multi-funded purchase order that split across three grant sources, handling a payroll adjustment for a union employee with multiple deduction types, and running a specific grant reimbursement report that federal auditors required quarterly.

The results shocked everyone who had watched the initial demonstrations.

Vendor A, whose presentation had seemed flawless, couldn't complete two of the three scenarios because their beautiful demo had been a carefully scripted performance following what vendors call the "golden path"—a route through the software that works perfectly as long as you never deviate from the script. When confronted with the messy reality of how the County actually operated, the system failed repeatedly, and the vendor's project manager, growing visibly uncomfortable as the evaluation proceeded, eventually acknowledged that the required functionality would involve significant custom development, adding what he estimated as $2–$3 million to the project cost.

Vendor B completed all three scenarios using standard configuration, revealing that their less-polished demonstration had actually been more honest, showing a real system handling real complexity rather than a theatrical performance designed to impress evaluators who wouldn't know the difference.

The county unanimously selected Vendor B, and the independent advisor's fee for the entire engagement proved to be less than a quarter of what that first change order from Vendor A would have cost. Two years later, the implementation is proceeding on schedule and on budget because the county invested in seeing reality rather than accepting the illusion that sophisticated vendor sales operations are designed to create.[77]

This case illustrates the central challenge every government faces during vendor selection: You're choosing a partner for a relationship that will last a decade or longer, in a market saturated with sophisticated sales operations that have studied exactly how to tell you what you want to hear, while obscuring the limitations you need to understand. Your only defense is a disciplined, evidence-based evaluation process that forces vendors to prove their capabilities against your specific reality rather than allowing them to perform carefully rehearsed demonstrations of idealized scenarios.

BEYOND THE DEMO: MAKING VENDORS PROVE IT

The traditional software demonstration represents theater rather than evaluation, with vendors investing enormous resources in creating demo environments specifically designed to highlight strengths,

---

77   Based on anonymized case studies from public sector IT consulting firms. The scenario is a common composite of real-world vendor selection "bake-offs."

while hiding limitations. The sales engineer presenting to you has delivered this demonstration hundreds of times and can navigate the golden path, while barely paying attention. They know exactly which features to emphasize and which screens to avoid, having anticipated your likely questions and rehearsed confident answers that sound responsive without committing to anything specific.

Making an informed decision requires taking control of the demonstration process and forcing vendors to prove their capabilities against the documented reality of your organization, which is precisely where your Phase Zero deliverables become invaluable. The Current State Assessment you developed during Phase Zero documented your most complex and problematic business processes, the Requirements Documentation captured what you actually need rather than generic industry requirements, and the AI-enabled requirements work identified the gaps and exceptions that standard demonstrations never address. Together, these artifacts form the foundation for scenario-based evaluation that strips away marketing polish to reveal genuine capability, or the lack of it.

Rather than allowing vendors to walk through their canned demonstrations, provide them with your documented current-state processes, your highest-priority pain points from Phase Zero, and your future-state vision. Next, ask them to demonstrate how their system solves your specific problems rather than hypothetical scenarios that conveniently align with their platform's strengths.

Successful scenario-based evaluations incorporate three essential elements that build directly on Phase Zero work. The first element involves creating "Day in the Life" scripts drawn from your most complex business processes, taking the three to five workflows that cause the most operational pain (i.e., the ones generating workarounds, complaints, and errors) and documenting them in enough detail for vendors to execute while maintaining sufficient open-endedness to reveal how their systems handle the exceptions and real-life edge cases. These scripts should come directly from your Phase Zero process documentation, capturing the actual complexity your staff navigates daily rather than simplified textbook scenarios that any vendor can handle.

The second element requires providing your actual data by giving each vendor a sanitized but realistic dataset extracted from your current systems. When vendors use their own perfectly curated demonstration data, they control the narrative completely. However, when they must use yours, the system's actual capabilities become visible in ways that sales presentations never reveal. In multiple postimplementation analyses, the U.S. Government Accountability Office has documented that failure to test systems with realistic data and workloads is one of the most common causes of implementation failure; nonetheless, organizations continue to accept vendor demonstrations using idealized data that bears no resemblance to operational reality.[78]

The third element involves assembling cross-functional evaluation teams that extend far beyond IT leadership and executives who will never interact with the system on a daily basis. Practitioners such as the accounts payable clerk who processes invoices, the payroll specialist who handles union deductions, and the budget analyst who produces quarterly reports for the board will see limitations that

---

78   United States Government Accountability Office, "IT DASHBOARD: Agencies Need to Improve Transparency and Accountability of Their Major Acquisitions," GAO-23-106277, September 2023.

a CIO will miss entirely because they know the exceptions, work-arounds, and edge cases that vendors' perfect demos never address. Their participation isn't a nice-to-have courtesy that demonstrates organizational inclusivity but an essential safeguard against selecting systems that shine in demos but falter in practice.

This evaluation approach strips away marketing glamour and forces direct comparison of how each system will actually function in your specific environment, which explains why vendors consistently resist scenario-based evaluation and push for their standard demonstration formats instead. That resistance tells you something important, and you should insist on your approach regardless of vendor objections.

THE ART OF THE REFERENCE CHECK

Every vendor will provide you with a carefully curated list of "reference customers" who will speak glowingly about their experience; however, these references prove nearly useless for making informed decisions because they represent the vendor's most successful implementations, often organizations that received extraordinary attention and support precisely because they were designated as reference accounts. The vendor has typically prepared these references before your call, sometimes even rehearsing key messages, so what you hear amounts to marketing dressed up as peer insight rather than genuine intelligence about how the vendor performs when things get difficult.

Understanding actual vendor performance requires going beyond the official list, which is another area where independent oversight and AI-enabled analysis create decisive advantage over organizations attempting to navigate selection without experienced guidance.

An experienced independent advisory firm maintains networks of government clients across dozens of jurisdictions and hundreds of implementations, which enables them to connect you with organizations similar to yours who are using the systems you're evaluating but who aren't on the vendor's approved reference list. These off-list references have no incentive to provide anything other than their honest experience, and they'll tell you about the implementation manager who disappeared mid-project to pursue a more lucrative opportunity, the functionality that was promised during sales but never actually delivered, and the change orders that doubled the budget after the contract was signed, and similar. This intelligence proves invaluable for predicting your own likely experience and remains essentially inaccessible without independent relationships that span multiple vendors and jurisdictions.

AI tools can supplement network-based intelligence with systematic analysis of the vast public information repositories that no human team could manually review—public meeting minutes where CIOs report on implementation progress to their boards, state auditor reports that document cost overruns and schedule delays, user group forums where practitioners discuss real-world challenges without vendor supervision, and news articles about troubled projects that vendors would prefer you never discover. AI can analyze thousands of these sources in a minute or less, building comprehensive pictures of vendor performance that would require months to assemble through traditional research methods.

During a recent state procurement, AI-assisted analysis surfaced public board meeting minutes from a county in another state where the CIO was detailing a series of missed deadlines and escalating scope disputes with a vendor who happened to be a finalist for the new contract.[79] This information had been publicly available for months but would never have emerged through traditional reference checking, and the state adjusted their evaluation criteria based on this intelligence before ultimately selecting a different vendor whose public record demonstrated more consistent delivery.

When you do speak with references, both from the vendor's curated list and from the off-list contacts your advisors identify, ask questions designed to penetrate the polished surface prepared references will present. Rather than asking whether they're satisfied with the system, which everyone on the vendor's list will affirm, ask about the biggest surprise they experienced during implementation, whether good or bad, because surprises reveal the gap between what was promised and what was delivered.

Ask them to describe the project manager the vendor assigned, specifically whether that person functioned as *a translator who helped bridge the gap* between technical and business perspectives or merely as *a transactor who processed paperwork without adding strategic value.* Ask how the vendor responded to the first major disagreement or the first significant change order request, because conflict response reveals organizational character more clearly than smooth sailing ever could. Ask to see their budget-to-actuals for the implementation and probe specifically about where the overruns occurred and why. Ask what they would negotiate differently in their contract if they could travel back in time with the knowledge they've accumulated since signing.

The answers to these probing questions reveal patterns that polished reference scripts never expose. For example, a vendor whose

---

79   News reports and public records are common sources for this kind of off-list vendor intelligence, and AI tools are used to rapidly scan and analyze such unstructured data sources.

references consistently describe surprise budget increases during the same implementation phase has a systemic estimation problem, while a vendor whose assigned project managers are repeatedly characterized as "responsive but not proactive" has a staffing model that prioritizes cost containment over client success. These patterns matter enormously for predicting whether your experience will resemble the success stories vendors want you to hear or the cautionary tales they hope you'll never discover.

THE STRATEGIC CHOICE: EMBEDDED AI VERSUS COMPOSABLE ERP

As you evaluate vendors, you confront a fundamental strategic decision that will shape your organization's technology trajectory for the next decade, a choice that has become increasingly consequential as AI capabilities have matured and integration technologies have evolved. The decision you make will determine not just your immediate implementation experience but your future flexibility, your ability to adopt emerging technologies, and your leverage in vendor relationships for years to come.

The choice lies between a single integrated ERP system with AI capabilities embedded throughout, typically from major vendors such as Oracle, SAP, or Workday, and a more flexible composable approach that combines core financial systems with best-of-breed solutions from multiple vendors connected through modern integration platforms.

The embedded approach offers compelling advantages that vendors emphasize heavily in their sales presentations. These benefits are genuine rather than mere marketing spin. When every module originates from the same vendor, data flows seamlessly between functions without complex integration work, and you negotiate and manage a single contract rather than juggling multiple vendor relationships. In addition, users encounter consistent interfaces across the entire platform rather than adapting to different design philosophies in different functional areas.

The AI capabilities these vendors embed are designed to work within their ecosystem, automating invoice processing in the financial module, suggesting career development paths in the HR module, and predicting budget variances based on historical spending patterns. Organizations with relatively standard operational needs and limited technical staff often find that this unified approach reduces complexity in ways that justify the constraints it imposes.

The embedded approach also carries substantial risks that vendor presentations consistently minimize or ignore entirely. Once you commit to a single vendor's ecosystem, switching becomes extraordinarily expensive and disruptive, creating leverage dynamics that shift dramatically in the vendor's favor following implementation. Your organization becomes tied to that vendor's development priorities and innovation pace. This means that if they're slow to develop capabilities in areas important to your operations, such as specialized grants management or public sector budgeting tools, your only options are to wait indefinitely, build expensive customizations that complicate future upgrades, or accept functionality that doesn't meet your needs.

The AI capabilities embedded in these platforms optimize for the vendor's strategic priorities rather than yours. It is not easy to substitute superior AI tools as they emerge from other sources because the embedded architecture assumes you'll use whatever the platform vendor provides.

The composable approach, which industry analysts at firms such as Gartner have increasingly advocated for organizations with complex needs and capable technical teams, embraces the opposite philosophy entirely.[80] Rather than accepting a single vendor's version of every function regardless of whether it represents the best available solution, you select the optimal tool for each business need, with your core financial system potentially coming from one vendor, your talent management platform from another, and your budgeting and forecasting capabilities from a third. Modern APIs and integration platforms have made connecting these systems far more manageable than was possible even 5 years ago, although the integration responsibility shifts from the vendor to your organization and its implementation partners.

This approach provides flexibility that embedded systems fundamentally cannot match, enabling you to adopt superior solutions as they emerge in any functional area without replacing your entire technology stack. It also protects your AI strategy from any single vendor's development roadmap and reduces the risk of catastrophic vendor lock-in because replacing one component of a composable architecture proves far less disruptive than extracting yourself from an integrated suite that touches every aspect of your operations.

The composable approach does demand more from your organization. You become responsible for ensuring that different systems communicate effectively, which requires technical sophistication

---

80   Gartner, Inc. has published numerous research papers and held webinars on the topic of "Composable ERP" and the shift away from monolithic systems, particularly for agile and large-scale organizations. See Gartner's annual "Magic Quadrant" reports for ERP systems.

and ongoing attention that some organizations lack the capacity or appetite to provide. You must manage multiple vendor relationships, multiple contracts, and multiple support channels rather than directing all inquiries to a single point of contact, and users may encounter different interfaces across systems, which increases training requirements and potentially creates friction in cross-functional processes. These costs are real and must be weighed honestly against the flexibility benefits rather than dismissed as manageable inconveniences.

The right choice depends entirely on your organization's specific circumstances—the complexity of your operations, your technical team's integration capabilities, your appetite for ongoing vendor management complexity, and your strategic priorities for the next decade. Your Phase Zero deliverables, particularly the AI Governance and Readiness Assessment and the Technology Landscape Analysis, should inform this decision substantially by revealing whether your organization possesses the technical maturity to manage a composable environment and whether your operational complexity genuinely requires best-of-breed solutions in multiple functional areas.

An independent advisor can help you analyze the tradeoffs based on your specific situation, but this remains a strategic decision that your executive leadership must own after thorough deliberation rather than a technical choice that can be delegated to IT staff and ratified without substantive engagement.

NEGOTIATION: WHERE CONTRACTS BECOME PROTECTION

After the demonstrations, reference checks, and strategic analysis, the final protection for your organization comes down to the contract itself. The contract is not a procurement formality to rush through once you've made your selection but rather the legally binding document that will govern your relationship for the next decade, with every clause carrying consequences that may not become apparent until disputes arise years after signing.

Vendor sales teams receive compensation tied to closing deals, while their legal teams are trained to minimize vendor risk by shifting it to you through carefully crafted language that sounds reasonable on first reading but creates significant exposure when problems emerge. The information asymmetry in these negotiations is enormous because your team negotiates major technology contracts occasionally, perhaps once every several years, while their team negotiates them constantly and has optimized every provision through countless iterations across hundreds of government clients.

Independent oversight proves its value most directly during contract negotiation because experienced advisors have negotiated hundreds of these agreements and know exactly where problematic provisions hide. They recognize the liability limitation clauses that sound like standard industry practice but are actually far more restrictive than market norms. They also spot the intellectual property provisions that would make future vendor transitions prohibitively expensive and understand which terms characterized by vendors as "industry standard" actually represent vendor-favorable language dressed up as universal convention. Without this experienced perspective, government negotiators consistently accept terms that more sophisticated buyers would reject, discovering the consequences only when trying to hold vendors accountable for commitments quietly undermined by the contract language.

Several negotiation principles consistently separate contracts that genuinely protect government interests from those that merely appear protective while containing carefully crafted loopholes.

Tying payments to performance outcomes rather than project milestones fundamentally changes vendor incentives in your favor. Standard vendor contracts structure payments around activities such as "software installed" or "training complete" the vendor controls entirely and milestones that can be achieved, and they trigger payment even when the system isn't actually working for your organization. Negotiating to tie significant payment portions to actual performance outcomes, such as 99% of payrolls processed accurately for three consecutive months, financial close completed within 5 business days, or user adoption reaching defined thresholds, aligns vendor financial incentives with your success rather than allowing them to collect payment for activities that don't translate into value.

Defining "go-live" in terms that matter to your organization prevents vendors from declaring premature victory and triggering payment milestones before you've achieved actual success. Vendors prefer to declare go-live success the moment the system becomes technically operational, starting warranty clocks and triggering payment schedules regardless of whether the organization can actually function effectively. Your contract should define go-live in business terms that reflect operational reality: The system is stable under production loads, users have demonstrated proficiency in essential functions, the key performance indicators established during Phase Zero are being met consistently, and the organization can operate without emergency vendor support for routine issues. The gap between technical go-live and business go-live often spans months, and the contract should reflect your definition of success rather than the vendor's.

Protecting your access to the people who sold you prevents the common bait-and-switch where experienced talent disappears after contract signing. Vendors present their best people during the sales process, the seasoned implementation managers, knowledgeable technical architects, and reassuring project executives who seem to understand your organization's specific challenges. Once the contract is signed, those impressive individuals often vanish to pursue the next sales opportunity and are replaced by junior staff who weren't part of the conversations that shaped your expectations. Your contract should specify that the key personnel you met during evaluation will lead your implementation through defined milestones and establish meaningful penalties or termination rights if they're reassigned without your explicit approval.

Building contract provisions that give your governance structure real authority transforms Phase Zero documentation from aspirational planning into enforceable requirements. The Executive Steering Committee charter you developed, the escalation procedures you defined, and the decision rights framework you established all need corresponding contract language that creates vendor obligations rather than mere expectations. A governance structure without contractual teeth remains merely advisory, allowing vendors to participate in governance meetings, while ignoring decisions they find inconvenient. Only explicit contract provisions prevent this erosion of the oversight framework you worked so hard to establish.

Negotiation is an endurance sport that vendors have trained for extensively across thousands of government contracts, and having an experienced independent advocate on your side of the table is the only reliable way to achieve agreements that genuinely protect your public investment rather than documents that appear protective, while containing carefully designed escape routes.

WHEN COOPERATIVE CONTRACTS MAKE SENSE

As mentioned in the procurement chapter, cooperative contracts offer significant advantages for certain engagements, and vendor selection is where strategic deployment of these vehicles delivers substantial value. If you engaged independent advisors through a cooperative contract during Phase Zero and procurement, that same relationship can provide continuity through vendor selection, with advisors who already understand your organization, your requirements, and your political dynamics guiding evaluation without the months of delay that a new competitive procurement would require.

For the major vendor selection itself, the strategic calculus somewhat differs because software and system integrator contracts involve significant customization, long-term commitment, and organization-specific fit. These warrant careful evaluation beyond what cooperative contract terms can guarantee. However, the independent oversight that guides that evaluation represents exactly the kind of professional service engagement where cooperative contracts excel, enabling you to access experienced advisors who provide vendor-neutral perspective without sacrificing months to procurement processes that add delay without adding value.

Organizations that navigate vendor selection most effectively deploy cooperative contracts strategically by engaging independent advisors quickly through these vehicles to guide rigorous evaluation of the software vendors and system integrators they'll partner with for the next decade, rather than allowing procurement timelines to delay access to the expertise that makes selection successful.

▎ KEY TAKEAWAYS

- Scenario-based evaluation using actual data and documented processes is the only reliable method for seeing past sophisticated vendor marketing. The demonstration a vendor wants to give you, rehearsed hundreds of times and optimized to highlight strengths while hiding limitations, reveals nothing meaningful about how their system will handle your specific operational complexity. The evaluation you design around your Phase Zero documentation, forcing vendors to execute real scenarios using your actual data with your front-line staff observing, reveals everything about the capacity of their platform to support your operations. Vendors who resist scenario-based evaluation or push back on using your data are communicating something important about what their polished demonstrations are designed to conceal.
- The choice between embedded suites and composable architectures shapes your organization's technology flexibility for the next decade and deserves executive-level

strategic deliberation rather than delegation as a technical decision that leadership ratifies without substantive engagement. Your Phase Zero deliverables, particularly the AI Governance and Readiness Assessment and Technology Landscape Analysis, should inform this choice by revealing your organization's technical maturity, integration capabilities, and operational complexity. Getting this decision wrong means either struggling with integration challenges you weren't prepared to manage or finding yourself locked into a vendor whose development priorities diverge from your needs as market conditions evolve.

- Contract negotiation represents your final line of defense and requires experienced independent representation to counter the information asymmetry that vendors exploit when government negotiators attempt to secure fair terms without specialized expertise. These are provisions that tie payments to business outcomes rather than technical milestones, define go-live in terms reflecting your operational reality rather than the vendor's preference, and protect your access to the talent whose competences convinced you to select this vendor. None of these protections appear in contracts unless you fight for them with advisors who have learned where the traps hide through hundreds of similar negotiations.

▌YOUR IMMEDIATE ACTION

Within the next 2 weeks, transform how your organization approaches vendor evaluation by converting preparation into evaluation instruments that force vendors to demonstrate genuine capability.

Begin by taking the three most complex business processes documented during your Phase Zero current state assessment (the workflows that generate the most complaints, workarounds and errors) and transforming them into detailed "Day in the Life" scripts that vendors must execute using your actual data. Communicate clearly to finalist vendors that their ability to complete these real-world scenarios will function as a primary evaluation criterion and observe carefully how vendors respond to this requirement. Those who push back on scenario-based evaluation or resist using your data are revealing important information about their confidence in their own systems.

Simultaneously, task your independent advisors with identifying and interviewing at least two off-list references for each finalist vendor. Going beyond the curated success stories vendors provide helps discover the factual experiences of organizations similar to yours who implemented these systems without the extraordinary support that designated reference accounts typically receive. Deploy the probing questions outlined in this chapter to move past surface satisfaction toward genuine understanding of implementation dynamics, budget variance patterns, and relationship quality when disputes arose.

Finally, schedule a dedicated workshop with your executive leadership and independent advisors to deliberate the embedded versus composable strategic choice, using your Phase Zero deliverables as the analytical foundation for a conversation that your leadership team must own rather than delegate. Your AI Governance and Readiness Assessment reveals whether your organization possesses the technical sophistication to manage composable complexity, while your Technology Landscape Analysis illuminates the integration challenges and opportunities that either approach would create. This decision commits your organization for a decade and deserves the executive attention and deliberation that commitment warrants.

Selecting a vendor represents the single most consequential decision in your entire modernization journey. Everything you built during Phase Zero—every requirement you documented, every governance structure you established, and every stakeholder relationship you cultivated—either pays off or gets compromised based on whether you identify a genuine partner or select a vendor whose sophisticated sales operation concealed limitations that your evaluation process should have revealed. A disciplined, independent, evidence-based selection process remains your only protection against the asymmetric information advantages that vendors have refined over decades of government contracting, and the investment in getting this decision right returns dividends for the entire decade of partnership that follows.

8

*Governance Reimagined:
Structures for Decision-Making
and Risk Management*

WHAT YOU'LL LEARN

The vendor you selected will soon begin implementation, and the contract you negotiated establishes the legal framework governing your relationship for the next decade. But contracts don't implement systems, vendor promises don't guarantee outcomes, and even the most carefully structured agreements can't prevent the problems that emerge when governance fails to maintain genuine oversight of complex, fast-moving projects.

What happens between contract signing and successful go-live depends almost entirely on governance—the structures, processes, and accountabilities that determine whether decisions get made when they need to be made, whether risks get identified and addressed before they become crises, and whether your organization actually changes how it operates rather than simply installing new software on top of old habits. This chapter illustrates:

- Why traditional project governance models consistently fail in modern ERP implementations despite the good intentions of everyone involved
- The three essential layers of successful governance and the critical Executive Sponsor role that makes the entire structure function
- How to integrate Organizational Change Management directly into governance rather than treating it as a parallel activity that coordinates loosely with project management
- How AI transforms risk management from backward-looking documentation into forward-looking intelligence that identifies problems, while intervention remains possible

The governance framework you establish in the coming weeks will determine whether your implementation joins the minority that succeed or the majority that struggle. Furthermore, the decisions you make now about structure and accountability will echo through every challenge your project encounters.

THE STEERING COMMITTEE THAT COULDN'T STEER

In 2021, a mid-sized Texas city launched an ambitious ERP modernization with every intention of doing things right, establishing what appeared to be a robust oversight structure following the best practices that consultants and industry publications had recommended for decades. They created a steering committee composed of every department head, the city manager, and two council members, assembling what seemed like comprehensive representation of organizational interests and sufficient political authority to clear any obstacle the project might encounter. The committee met monthly to review a 100-page status report prepared by the vendor's project manager, dedicating time that busy executives carved from demanding schedules to fulfill their governance responsibilities.

For the first 6 months, every indicator showed green and every milestone appeared on track, with the vendor's polished presentations describing steady progress and the committee members nodding along before returning to the operational demands that consumed most of their attention. The executives would listen to the 20-minute vendor presentation, ask a few questions that the project manager answered with reassuring confidence, approve the previous meeting's minutes, and adjourn feeling that they had discharged their oversight obligations appropriately. They were governing, they were steering, and the illusion of control felt complete enough that no one questioned whether the green indicators reflected project reality or merely vendor optimism.

Then, in Month 7, the project manager announced what he characterized as a "minor schedule adjustment," explaining that several key modules would be delivered 6 months later than originally planned and that the budget would require a 20% contingency increase to "account for unforeseen complexities" detected during detailed configuration work. When the city manager pressed for a root cause analysis, the meeting descended into a circular firing squad of finger-pointing that revealed how little anyone actually understood about the project's true condition. The vendor blamed the city's subject matter experts for being slow to make decisions and unavailable for critical working sessions that configuration required. The IT Department blamed the vendor for not providing clear options and adequate documentation that would have enabled faster decisions. The Finance Department blamed the HR Department for requesting a change that created cascading impacts across multiple modules. Everyone had someone else to blame, and no one could explain how a project that had shown green for 6 consecutive months had suddenly turned catastrophically red.

The committee found itself blindsided and paralyzed, unable to determine what had actually happened, who bore responsibility, or what should be done about it. The truth, which emerged painfully over subsequent weeks of investigation, was that the project had been in serious trouble for months, with warning signs visible to anyone paying close attention to the right indicators. However, the governance structure had been designed in ways that concealed problems rather than revealing them. The vendor-supplied status report functioned as a marketing document rather than an honest assessment, written by people whose incentives favored optimistic presentation over transparent reporting because their continued engagement depended on the city's confidence in project progress. The monthly meeting cadence was far too infrequent to maintain genuine situational awareness of a complex implementation where conditions could change dramatically in the weeks between sessions. Most critically, with no single point of executive authority empowered to make difficult decisions and enforce accountability, every tough choice had been deferred because no one wanted to create conflict, debated into gridlock because consensus proved impossible among stakeholders with competing interests, or delegated back down to the project team until small problems festered into crises that could no longer be hidden.

This pattern represents the classic failure mode of traditional project governance, creating an elaborate illusion of oversight, while allowing projects to drift inexorably toward disaster. Effective governance isn't a monthly meeting where executives listen to vendor presentations and approve minutes. It's a dynamic, multilayered system for decision-making, risk management, and organizational change that maintains genuine visibility into project reality and concentrates authority in roles empowered to act decisively on what that visibility reveals.

BEYOND THE ORG CHART: A MODERN GOVERNANCE FRAMEWORK

Successful ERP governance cannot function as a single committee attempting to address strategic, tactical, and operational concerns in the same periodic meeting, because the different types of decisions and oversight activities required at each level demand different participants, information, meeting cadences, and decision-making processes. The governance structure your implementation requires must operate across three distinct tiers, each designed to manage specific aspects of the project at the appropriate level of detail and authority while maintaining clear connections to the other layers.

This framework builds directly on the governance planning you initiated during Phase Zero, transforming the preliminary structures documented in your Governance Structure and Decision Rights deliverable into operational reality that begins with implementation. The stakeholder analysis completed during that phase, the political dynamics mapped across departments and leadership levels, and the decision-making patterns observed throughout your organization all inform you how to staff these governance bodies, design their operating procedures, and establish the escalation paths that connect them.

THE EXECUTIVE STEERING COMMITTEE: PROTECTING THE VISION

The strategic layer of your governance structure exists to protect the *why* of your implementation, ensuring that the business case and transformation vision you developed during Phase Zero remain the north star guiding every significant decision throughout the project rather than eroding gradually under the pressure of competing interests and short-term convenience. This committee doesn't manage everyday implementation activities, resolve configuration disputes between workstreams, or approve individual change requests. It guards the strategic intent that justifies the entire investment and intervenes when that intent faces compromise.

Composition matters enormously at this level, and the instinct toward broad representation that feels politically necessary typically undermines the committee's ability to function effectively. The Texas city's committee included every department head precisely because no one wanted to exclude important stakeholders and suffer the political consequences of perceived marginalization. However, that inclusive approach created a body too large for genuine deliberation and too politically complex for decisive action when stakeholders disagreed. Your Executive Steering Committee should remain small and empowered, typically including the City Manager or County Administrator, the Chief Financial Officer, the Chief Human Resources Officer, the Chief Information Officer, and perhaps one or two other executives whose functions are most directly affected by the implementation. Everyone else receives regular communication about project status and maintains channels for providing input on decisions affecting their areas. However, decision-making authority concentrates in a group small enough to actually make decisions when time pressure demands action.

Meeting frequency should balance the need for regular oversight against the reality that executives have demanding responsibilities beyond any single project, with monthly meetings of approximately 1 hour serving as the standard cadence supplemented by the ability to convene emergency sessions on 48 hours' notice when circumstances require urgent executive attention. The committee's mandate centers on guarding the transformation vision against the scope creep and political pressure that inevitably emerge as implementation proceeds, clearing major roadblocks the project team cannot resolve without executive intervention, championing the change effort across the organization in ways that signal leadership commitment, and owning ultimate accountability for budget, timeline, and outcome achievement.

But even a well-composed, appropriately empowered steering committee can become a forum for debate rather than decision when difficult choices create conflict among members with different departmental interests and competing priorities. Committees deliberate, build consensus, and balance perspectives. Yet committees struggle to decide when deciding means some stakeholders don't get what they want. This reality explains why the single most important role in the entire governance structure, and arguably in the entire implementation, is the Executive Sponsor.

THE EXECUTIVE SPONSOR: THE DECIDER YOUR PROJECT REQUIRES

The Executive Sponsor serves as the ultimate owner of the project's success, the individual whose accountability for outcomes is personal and undivided rather than distributed across a committee where responsibility diffuses until no one truly owns anything. This role is emphatically not a ceremonial title assigned to whoever has bandwidth or draws the short straw in a leadership meeting. This is the person who possesses both the positional authority and the accumulated political capital to make unpopular but necessary decisions in the best interest of the entire organization. They serve as the tie-breaker when the steering committee cannot reach consensus and as the enforcer when stakeholders resist changes they've nominally agreed to support.

Consider the specific decisions that derail implementations when no one has authority to resolve them decisively. When the Finance Director insists on a custom report that conflicts with the project's mandate to minimize customization and maximize use of standard functionality, someone must make the final call about whether that customization is truly essential for statutory compliance or represents the kind of scope creep that inflates budgets, extends timelines, and complicates future upgrades. When the Public Works department resists adopting standardized processes because their current approach feels more comfortable despite creating inefficiencies that the new system was intended to eliminate, someone must sit with that department head and explain that standardization is no longer optional, that the decision has been made at the organizational level, and that continued resistance will not change the outcome but damage the resister's standing with leadership. When the project team needs a decision by Thursday to maintain schedule but the relevant stakeholders cannot agree despite multiple attempts at resolution, someone must break the tie decisively rather than allowing the decision to drift until delay creates its own consequences that constrain future options.

A successful Executive Sponsor must possess several characteristics that cannot be compromised without undermining the role's effectiveness. Positional authority means they hold a role high enough in the organization (typically City or County Manager, Chief Financial Officer, or Deputy CAO) that their decisions carry weight no one can reasonably challenge and their directives translate into action rather than suggestions that stakeholders can quietly ignore. Political capital means they command respect across the organization through demonstrated competence and accumulated relationships, possessing the informal influence to persuade reluctant stakeholders through conversation, while also being able to defend the project against political attacks that inevitably emerge when change creates discomfort for people with connections to elected officials or other power centers. An organization-wide view means their primary loyalty is to the success of the entire enterprise rather than any single department, which enables them to make decisions that may disadvantage one function for the benefit of the whole without being perceived as favoring their home department's interests. And willingness to be "the bad guy" means they will say no to powerful department heads even when doing so creates friction, hold people accountable for commitments they've made even when accountability feels uncomfortable, and accept the interpersonal tension that decisive leadership inevitably creates in organizations where consensus has been the norm.

Without a strong, active Executive Sponsor who exercises these authorities regularly rather than reserving them for extreme circumstances, the Steering Committee will inevitably descend into consensus-driven gridlock where the path of least resistance wins out over the path to success. Difficult decisions get deferred because no one wants to create the conflict the resolution requires. Scope expands incrementally because saying yes to each individual request feels easier than saying no and dealing with the pushback. Timelines slip because holding people accountable for missed commitments feels harsh in cultures that prize harmony. Budgets swell because each addition seems small in isolation and no one aggregates the cumulative effect until the numbers become undeniable. And if the project struggles or eventually fails, everyone shares blame diffusely because no one was truly in charge, no one's personal accountability was on the line, and no one had both the authority and the incentive to prevent the drift that led to disaster.

THE PROJECT MANAGEMENT OFFICE: YOUR IMPLEMENTATION'S NERVE CENTER

The tactical layer of your governance structure operates through the Project Management Office, which functions as the nerve center coordinating all implementation activities, tracking progress against plans with sufficient granularity to identify emerging problems, surfacing risks before they materialize as crises, and ensuring that the Steering Committee and Executive Sponsor receive accurate information about project reality rather than optimistic presentations designed to avoid difficult conversations.

The fundamental choice at this layer involves whether you staff the PMO with internal resources or establish an Independent Verification and Validation model where your independent advisory partner operates the PMO function. This decision connects directly to the independence themes that have run throughout your Phase Zero work, your procurement strategy, and your vendor selection process, representing the operational manifestation of the structural protection you've been building.

A self-managed PMO can function effectively when your organization possesses a deep bench of project managers with recent, successful experience leading large-scale ERP implementations in government environments. However, this internal model constantly battles pressures that compromise the objectivity essential for effective oversight. As the Government Accountability Office has repeatedly documented in analyses of federal IT project failures, lack of independent oversight consistently emerges as a primary contributor to projects that run over budget, behind schedule, or fail to deliver intended benefits.[81] An internal project manager reporting to internal leadership faces structural pressures that even the most professional

---

81   United States Government Accountability Office, "Information Technology: A Framework for Assessing and Improving Enterprise Architecture Management (Version 2.0)," GAO-10-846G, September 2010.

and well-intentioned individuals struggle to overcome: They may hesitate to deliver unwelcome news about project status to executives who control their career advancement, struggle to push back against powerful department heads whose cooperation they need for success in their current role and future opportunities, or unconsciously shade assessments in optimistic directions because their professional reputation becomes intertwined with project success in ways that make objective reporting psychologically difficult.

The Independent Verification and Validation Model addresses these structural challenges by placing PMO functions under your independent advisory partner, who is insulated from your internal politics and maintains no financial relationship with the software vendor or system integrator. Their only metric for success is your success, which creates incentives aligned entirely with accurate reporting and effective project execution rather than the mixed incentives that internal staff inevitably face. They provide unvarnished, data-driven reports to the Steering Committee that make it impossible for problems to hide behind optimistic presentations or bureaucratic obfuscation. Their professional reputation and future business depend on the success of your implementation rather than on maintaining comfortable relationships with internal stakeholders who might prefer not to hear difficult truths. This is the model Monroe County employed throughout their implementation, and it represents the single most effective insurance policy against the kind of surprise that blindsided the Texas city's steering committee when a project everyone believed was progressing smoothly suddenly revealed months of hidden problems.

OPERATIONAL WORKSTREAMS: WHERE IMPLEMENTATION ACTUALLY HAPPENS

The operational layer is where actual implementation work occurs, with teams of your subject matter experts working alongside the vendor's consultants to configure the system according to requirements, validate that configurations meet the needs documented during Phase Zero, develop procedures for new processes that the system will enable, and prepare the organization to operate differently than it has in the past.

Staffing these workstreams with your best people is absolutely nonnegotiable, despite the operational disruption it creates and the temptation to assign whoever happens to be available without creating gaps in current operations. The people who are available are typically available for a reason, and the reason is rarely that they're your strongest performers whom you've been holding in reserve for exactly this opportunity. Successful implementations require your stars and informal leaders—the people who truly understand how work gets done and who command the respect of their peers in ways that give their endorsement of new processes credibility that no executive mandate can substitute for.

Research from firms such as McKinsey demonstrates direct correlation between the quality of business-side team members assigned to implementation workstreams and ultimate implementation success, which makes intuitive sense when you consider that these are the people making thousands of configuration decisions that will shape how the system actually functions for years to come.[82] A marginal

---

82   Michael Bloch, Sven Blumberg, and Jürgen Laartz, "Delivering large-scale IT projects on time, on budget, and on value," McKinsey & Company, October 2012.

performer who doesn't fully understand current operations will make configuration choices that seem reasonable in the abstract but create problems when the system encounters the real-world complexity that your best people would have anticipated. An informal leader whose peers respect their judgment will build support for new processes in ways that ease adoption, while someone without that standing will struggle to overcome resistance regardless of how technically correct their decisions might be.

Indeed, backfilling positions during implementation creates expense and operational disruption that budget-conscious managers instinctively resist. Pulling your best people from their regular responsibilities does create short-term challenges that require creativity and flexibility to manage. However, these costs pale in comparison to the expense of failed implementations taking place when workstreams are staffed with people who lack the knowledge, credibility, or decision-making authority; who make configuration decisions based on incomplete understanding; and who cannot build the peer support that adoption ultimately requires. Your Phase Zero stakeholder analysis identified these key individuals across every functional area. Now you must actually assign them to implementation workstreams despite the organizational pain their temporary absence from regular duties creates.

INTEGRATING ORGANIZATIONAL CHANGE MANAGEMENT

Where does Organizational Change Management fit within this governance framework? In failed implementations, OCM operates as an afterthought, a communications function that sends newsletters about project milestones and schedules training sessions without genuine integration into project decision-making or influence over how decisions account for their change impact. In successful implementations, OCM serves as the connective tissue binding the entire governance structure together, ensuring that every significant decision considers its effect on organizational readiness and that every phase gate incorporates assessment of whether the organization is truly prepared to absorb the changes that technical progress has made possible.

Your independent advisory partner should lead the OCM effort as an integral component of the PMO rather than a separate workstream that coordinates loosely with project management through occasional meetings and shared status reports. This integration means the OCM lead attends every Steering Committee meeting, providing perspective on how proposed decisions will affect the organization's readiness to adopt new systems and processes and flagging concerns that purely technical perspectives might miss. It means that change readiness assessments function as formal gateways that must be passed before the project advances to subsequent phases, which prevents the common pattern where technical milestones are achieved on schedule, while the organization remains unprepared to actually use what's been built. In addition, it helps avoid situations where go-lives meet technical criteria but fail to deliver intended benefits because users aren't ready to work differently.

This approach moves communication beyond generic project updates announcing milestones and celebrating progress. Instead, specific roles and functions are targeted with messages that address the fundamental question every employee asks when confronting organizational change: "What does this mean for me personally, and why should I support it rather than resist it?" The accounts payable clerk doesn't need to know about overall project progress—they need to know specifically how their daily work will change, what new skills they'll need to develop, what will become easier, and what support they'll receive during the transition. The department head doesn't need celebratory messaging about enterprise transformation—they need honest assessment of how their team's workflows will change and what their role is in preparing their staff for those changes.

The stakeholder engagement strategy you developed during Phase Zero established the foundation for this OCM integration by mapping influence relationships across the organization, identifying likely sources of resistance and the concerns driving it, and defining engagement approaches tailored to different stakeholder segments based on their specific interests and concerns. Your governance structure must now operationalize that strategy through regular assessment of stakeholder sentiment as implementation proceeds, proactive intervention when resistance emerges, and visible celebration of adoption successes that reinforce the message that change is possible and beneficial.

AI-POWERED RISK MANAGEMENT: FROM DOCUMENTATION TO INTELLIGENCE

Traditional risk management relies on risk registers that project managers update periodically, documenting known risks, assigning probability and impact ratings based on subjective assessment, and defining mitigation strategies that may or may not be executed with the discipline their importance warrants. This approach treats risk management as an administrative task to be completed rather than an intelligence function that actively protects the project, thus creating comprehensive documentation that captures what was known at the last update rather than revealing what's actually happening now as conditions evolve.

AI transforms risk management from reactive administration into proactive intelligence that identifies emerging problems before they manifest as crises, connecting directly to the AI governance capabilities assessed during Phase Zero and the AI-enabled approaches to requirements and procurement discussed in earlier chapters.

Consider the difference between traditional and AI-enabled governance when the Steering Committee meets to assess project status. In the traditional model, the vendor presents a status report showing green indicators across all workstreams, the PMO reports yellow status on the testing workstream based on manually compiled information from the previous week, and the committee engages in vague discussion about "getting testing back on track" without clear understanding of what's actually wrong or what specific actions would address the problem. The Executive Sponsor might ask questions, but the information available doesn't support the precise, actionable answers that effective intervention requires. Decisions drift because the governance system generates insufficient intelligence to drive decisive action.

In the AI-enabled model pioneered by organizations such as the City of San Diego, the independent PMO displays a governance dashboard that continuously monitors what practitioners call the project's "digital exhaust," the emails, support tickets, code commits, meeting patterns, and communication dynamics that reveal actual project conditions rather than the curated presentations that traditional reporting provides.[83] The dashboard reveals that while the vendor's official report shows green status, sentiment analysis of project team communications identifies a sharp decline in positive sentiment within the testing workstream over the past 2 weeks. Quantitative metrics show that defect closure velocity has decreased by 30% while new defect discovery has accelerated, suggesting the team is falling behind rather than converging toward completion. Predictive models trained on patterns from thousands of similar implementations indicate that these specific signals, when they co-occur at this project phase, create a 70% probability of a 60-day go-live delay if root causes aren't addressed in the next 10 business days.

Armed with this intelligence, the Executive Sponsor can ask fundamentally different questions that drive fundamentally different outcomes: "What is the specific root cause problem on the testing team, who is accountable for resolving it, and what is the concrete plan to address it by next Friday with measurable indicators that we're making progress?" This is the difference between steering and drifting—between governance that maintains genuine control over project trajectory and that that creates an illusion of oversight while problems compound beneath the surface until they become too large to hide.

The AI Governance and Readiness Assessment you completed during Phase Zero identified your organization's capacity to deploy these AI-enabled oversight tools, evaluating both the technical infrastructure required and the organizational readiness to act on

---

83   City of San Diego, IT Department, "ERP Governance Dashboard Initiative," 2024.

the intelligence these systems provide. Your vendor selection process evaluated whether prospective partners could support the transparency that AI-enabled monitoring requires, including contractual provisions ensuring access to the project data streams that make this monitoring possible. Now your governance structure must operationalize these capabilities, establishing dashboards and monitoring systems that provide continuous visibility into project dynamics rather than waiting for periodic reports that may arrive too late to enable effective intervention.

❚ KEY TAKEAWAYS

- Governance requires a Decider whose personal accountability for outcomes creates the urgency and authority that committees inherently lack. Steering committees serve essential functions by building consensus, balancing perspectives, and ensuring that diverse interests receive consideration, but a single empowered Executive Sponsor with genuine positional authority, accumulated political capital, and an organization-wide perspective remains essential to break ties when stakeholders cannot agree, enforce decisions when departments resist changes they've nominally accepted, and hold the entire project accountable to the transformation vision established during Phase Zero. Without someone filling this role who is willing to exercise its authorities even when doing so creates friction, governance structures devolve into consensus-seeking forums where the path of least resistance prevails over the path to success, where difficult decisions get deferred until delay forecloses options, and where no one owns outcomes because responsibility has been distributed until it disappears.

- Independence transforms the Project Management Office from a coordination function into a truth-telling function whose value lies precisely in its insulation from the pressures that compromise internal reporting. An Independent Verification and Validation model places your PMO under advisors whose only success metric is your success, who maintain no relationships with vendors that might color their assessments, and who can deliver unwelcome news without fearing career consequences from executives who might prefer not to hear it. This independence creates conditions where accurate reporting reaches the Steering Committee and Executive Sponsor regardless of how uncomfortable that accuracy makes stakeholders whose performance is being assessed, enabling intervention while problems remain manageable rather than discovering issues only after they've grown too large to address without significant cost and delay.
- AI elevates risk management from backward-looking documentation that captures what was known at the last update to forward-looking intelligence that reveals what's happening now and predicts what's likely to happen next. By monitoring the digital patterns that reveal actual project dynamics rather than relying on manually compiled status reports that arrive days or weeks after conditions change, AI-enabled governance dashboards identify emerging problems while intervention remains possible, trans-

forming potential crises into manageable issues that the Executive Sponsor can address decisively. This capability requires the AI governance foundation you established during Phase Zero and the transparency commitments you negotiated into your vendor contracts, demonstrating how preparation in earlier phases enables capabilities that would otherwise be impossible.

▌ YOUR IMMEDIATE ACTION

Within the next two weeks, establish the governance foundation that will guide your implementation through every challenge and decision point ahead, beginning with the role that determines whether the entire structure functions or fails.

Start by formally appointing your Executive Sponsor before any other governance element takes shape, because this role defines accountability for the entire implementation and influences every subsequent decision about structure, process, and authority. Your senior leadership must officially designate someone who possesses the positional authority, political capital, organizational perspective, and willingness to make difficult decisions that the role demands, someone whose personal reputation will be tied to implementation outcomes in ways that create genuine accountability rather than distributed responsibility that no one truly owns. That person's first official act should be signing the Governance Charter that formalizes their accountability and authorities, making explicit the expectations that the organization has for this role and the support that the organization commits to providing.

Then draft your three-tiered Governance Charter, defining with specificity the membership criteria, meeting frequency, mandates, and decision rights for your Executive Steering Committee, your PMO structure including whether you will employ an independent IV&V model, and your key Operational Workstreams. This charter should translate the preliminary governance planning from your Phase Zero deliverables into operational procedures that take effect when implementation begins, establishing how decisions will be made at each level, how conflicts will be escalated when the appropriate level cannot resolve them, how information will flow between layers, and how accountability will be enforced when stakeholders don't meet commitments they've made. The clarity you establish now prevents the ambiguity that allows problems to fester and decisions to drift.

Finally, task your CIO with researching and demonstrating AI-powered project monitoring and risk management tools that could provide the continuous visibility your governance structure requires to function effectively. Seeing these capabilities in action fundamentally changes how leadership thinks about project oversight, revealing the gap between traditional periodic reporting that may arrive too late for effective intervention and the continuous intelligence that AI-enabled governance provides. The AI Governance and Readiness Assessment from your Phase Zero work should inform this evaluation, identifying which tools align with your organization's technical capabilities, data infrastructure, and tolerance for the transparency that effective AI monitoring requires.

The governance structure you establish in these coming weeks will face every challenge your implementation encounters, from vendor disputes to stakeholder resistance to unexpected technical complications to the inevitable moments when project reality diverges from project plans. A well-designed structure led by a strong Executive Sponsor who exercises authority decisively, managed by independent advisors whose only interest is your success, and powered by AI tools that reveal actual project dynamics rather than curated presentations creates the conditions where challenges become manageable problems rather than cascading crises. The Texas city's steering committee discovered too late that their governance structure had been designed in ways that guaranteed failure despite everyone's good intentions; the structure you build must be designed from the start to succeed.

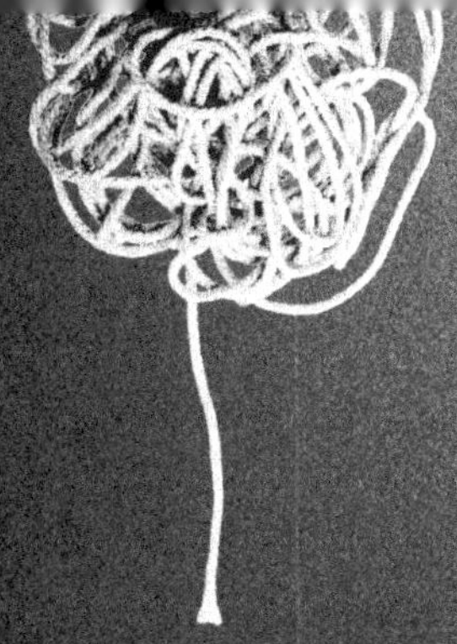

PART IV

The Human Dimensions

This part covers what happens after the structure is in place: organizational change management, stakeholder alignment, and the advisory relationships that determine whether governance functions as intended or becomes bureaucratic theater. It also examines how artificial intelligence is reshaping the modernization landscape in ways that make independence more essential rather than less, and it closes with a five-phase roadmap that synthesizes everything into a practical plan for breaking free. If your implementation is technically sound but struggling with adoption, or if you are weighing where AI fits into your next set of decisions, start here.

9

The Human Element:
Winning Hearts and Minds

WHAT YOU'LL LEARN

The governance structure established in the previous chapter created the decision-making framework your implementation requires, with an Executive Sponsor empowered to resolve conflicts, an independent PMO providing unvarnished visibility into project status, and operational workstreams staffed with your best people making the configuration decisions that will shape how your new system actually functions. But governance structures don't change how people work, and technical configurations don't determine whether employees embrace new processes or resist them until resistance becomes the default that no amount of executive mandate can overcome.

The difference between implementations that deliver their intended benefits and those that install new software without actually transforming operations is almost entirely in how organizations manage the human dimensions of change. This chapter illustrates

- Why the communication strategies that feel natural and logical are almost guaranteed to fail
- The practical tactics that actually secure stakeholder buy-in rather than merely informing stakeholders that change is coming
- How to identify and constructively address the four distinct types of resistance that emerge in every major implementation
- How AI-enabled training approaches are revolutionizing user adoption in ways that compress the time to value from years to months

The stakeholder engagement strategy developed during Phase Zero establishes the foundation; now you must execute that strategy with the discipline and persistence that genuine organizational change requires.

THE EMAIL NO ONE READ

In 2022, a large school district on the West Coast reached the 6-month mark before go-live of its new HR and payroll system with every reason to feel confident about their progress. The implementation was technically on track, with configuration work proceeding according to schedule and testing revealing manageable issues rather than fundamental problems. The governance structure discussed in the previous chapter was functioning effectively, with an engaged Executive Sponsor, regular steering committee meetings that addressed issues before they became crises, and workstreams staffed with knowledgeable subject matter experts who understood the district's operational complexity. The project team, recognizing that technical progress meant nothing if employees weren't prepared to work differently, launched what they believed was a comprehensive communication plan designed to prepare the workforce for the transition ahead.

Every Friday, a professionally designed email newsletter reached all 7,000 district employees, featuring project updates written in accessible language, explanations of new system features presented with engaging graphics, and a "Meet the Team" section that humanized the project by introducing the people behind it. The project's SharePoint site accumulated FAQs addressing common questions, process documents explaining how workflows would change, and training schedules that employees could reference as go-live approached. The OCM team tracked their metrics diligently: emails sent, SharePoint page views, newsletter open rates. By every measure they were tracking, they were communicating effectively, checking all the boxes that their change management methodology prescribed.

Then, 3 months before go-live, a change readiness survey was conducted, designed to assess whether the organization was actually prepared for the transition, and the results revealed a chasm between what the project team believed they had accomplished and what employees had actually absorbed. Over 60% of respondents reported feeling "uninformed" or "very uninformed" about the project, despite 6 months of weekly newsletters reaching their inboxes. A staggering 75% indicated they didn't understand how the new system would affect their daily responsibilities, the specific question that mattered most for adoption readiness. Perhaps most alarmingly, the survey's open-ended responses revealed that rumors were spreading through the district's informal communication channels: The new system was designed to eliminate positions and reduce headcount, it was too complicated for normal people to use, and it would inevitably mess up everyone's paychecks during the transition. These rumors had spread unchecked because the project's official communications weren't reaching the people they were intended to inform.

The project director's reaction captured the confusion that many teams experience when confronted with similar gaps between communication effort and communication impact: "But we've been sending emails for months! How can people say they're uninformed?"

Their mistake was one that well-intentioned project teams make repeatedly: They confused broadcasting information with genuine communication, measuring their success by what they sent rather than what was received, understood, and believed. They were so focused on pushing messages outward through official channels that they never verified whether those messages were penetrating the informal networks where employees actually formed their opinions about organizational changes. They failed to grasp the fundamental law of organizational change that determines outcomes regardless of technical excellence: A project's success depends not on the quality of the system being implemented but on the willingness of people to use it in ways that capture its intended benefits.

This is the domain of Organizational Change Management, understood not as a soft, optional activity that supplements the real work of technical implementation but as a hard-nosed, disciplined strategy for managing the human dynamics that determine whether technology investments generate returns or become expensive infrastructure that nobody uses effectively.

BEYOND NEWSLETTERS: WHAT ACTUALLY WORKS

Effective OCM bears little resemblance to the newsletter-and-Share-Point approach that failed the school district, operating instead as a proactive, data-driven effort to build genuine buy-in among stakeholders whose support matters, manage resistance constructively rather than hoping it dissipates on its own, and prepare the workforce for a new operational reality that may differ substantially from the habits they've developed over years or decades. This work connects directly to the Stakeholder Analysis and Engagement Strategy you developed during Phase Zero, operationalizing the insights you gathered about influence patterns, likely resistance sources, and communication preferences into sustained engagement that continues throughout implementation.

SECURING GENUINE BUY-IN

Buy-in is not created through a single all-hands meeting where executives announce the project and encourage everyone to embrace change, nor does it emerge automatically from well-designed newsletters that employees delete without reading. Genuine buy-in develops through deliberate, sustained, and precisely targeted engagement that addresses what employees actually care about rather than what project teams want to communicate.

The principle that should guide every communication your project produces is deceptively simple but consistently violated: Every message must answer the question that every employee asks when confronted with organizational change, which practitioners refer to as "What's In It For Me?" This question isn't selfishness or resistance to organizational priorities but the natural human response to uncertainty about how change will affect one's daily experience, job security, and professional identity. A generic message about "improving organizational efficiency" or "modernizing our technology infrastructure" means nothing to a payroll clerk who processes the same transactions every day and worries primarily about whether the new system will make their job harder, threaten their position, or disrupt the routines they've developed over years of experience. A specific message about "eliminating the manual data entry that currently takes you until Friday each week, so you can complete payroll processing by Wednesday and have time for the analytical work you've told us you'd rather be doing" is an effective message that gets remembered and repeated because it addresses what the recipient actually cares about.

Transforming this principle into practice requires the kind of granular understanding of different employee groups that your Phase Zero stakeholder analysis should have developed. Each major role category experiences change differently, worries about different consequences, and needs to hear different messages about why this change benefits them. The investment in understanding these differences before implementation began now pays dividends through communications that actually resonate rather than generic project updates that get lost in inbox clutter.

Equally important is recognizing that you cannot communicate effectively with thousands of employees through messages emanating from a central project office, regardless of how well-crafted those messages might be. Employees don't form their opinions about organizational changes based on official communications—they form opinions through conversations with colleagues they trust, observations about how leaders are behaving, and interpretations of organizational signals that may have nothing to do with what the project team intended to convey. To influence this informal communication network, you must build what practitioners call a Change Champion network: respected informal leaders from every level and every corner of the organization who carry accurate information and consistent messages to their peers with credibility that no project team communication can match.

These Change Champions are not necessarily managers or formal leaders. They're the people others turn to when they want to know "what's really going on," the informal connectors whose opinions shape how their colleagues interpret organizational events. Research by McKinsey and other firms studying organizational change consistently shows that leveraging these informal leaders represents one of the most effective strategies for driving successful adoption because messages delivered by trusted peers penetrate resistance that official communications cannot overcome.[84] Your job is to identify these individuals, equip them with accurate information and clear talking points that address the concerns their peers are likely to raise, and maintain ongoing dialogue that keeps them informed as the project evolves. They become an extension of your OCM function, carrying messages into conversations that your project team will never be part of and countering rumors before those rumors crystallize into resistance.

---

84   McKinsey & Company, "The people power of transformations," May 2017.

The Executive Sponsor whose critical role was discussed in the previous chapter serves an equally essential function in the communication effort, but that function requires visible, active engagement rather than passive endorsement. Employees need to see the sponsor, hear directly from them, and believe that they are genuinely committed to this change rather than merely lending their name to an initiative that others will actually lead. The sponsor must be the lead voice in town hall meetings where employees can ask questions and observe how leadership responds to concerns. Their name and face must appear on key communications in ways that signal personal ownership rather than bureaucratic approval. Finally, they must be seen using the new system once it's available and championing its benefits based on their own experience. When employees observe that the most senior leaders are personally invested in the change, that observation influences their own willingness to engage in ways that no amount of project team messaging can achieve.

MANAGING THE FOUR TYPES OF RESISTANCE

Resistance to organizational change is not a character flaw, a sign of poor attitude, or evidence that employees don't care about organizational success. It is a natural human reaction to uncertainty, disruption of established routines, and threats to professional identity that deserves understanding rather than dismissal. Your job is not to crush resistance through executive mandate or shame resisters into compliance, but to understand the sources driving resistance in each case and address those sources constructively so that resisters become supporters rather than obstacles.

In my experience guiding organizations through major implementations, resistance typically emerges from four distinct sources that require different responses, and correctly diagnosing which type you're facing determines whether your intervention helps or makes things worse.

The Skeptic has seen transformational projects come and go throughout their career, and most of those projects failed to deliver what they promised despite confident leadership assurances that this time would be different. Their resistance is rooted in legitimate experience that deserves respect rather than dismissal, because they've watched previous initiatives consume organizational energy, disrupt operations, and ultimately fade away without achieving their objectives. Handling Skeptics effectively requires acknowledging past failures honestly rather than pretending this organization's history doesn't exist, followed by showing them specifically how this project differs from the failures they remember. When a Skeptic says, "We tried something like this 5 years ago and it was a disaster," the wrong response is defensive denial; the right response acknowledges their concern and explains concretely what's different this time, perhaps noting, "You're right that the last project failed because we didn't have independent oversight and let the vendor control everything. Here's how our IV&V model is protecting us now, and here's the evidence that it's working." Skeptics who feel their experience is respected and their concerns are addressed with evidence rather than reassurance often become your strongest champions precisely because their conversion carries credibility with other skeptics who share their concerns.

 **The Expert** derives significant professional identity from deep knowledge of current systems, serving as the go-to person everyone consults when they need to understand how things really work beneath the official procedures. The new system threatens this status in fundamental ways because expertise accumulated over years becomes less valuable when the systems that expertise relates to are replaced. What looks like resistance to change is often actually fear of losing the status and security that expertise provides. Handling Experts effectively requires making them part of the solution rather than victims of progress, designating them as Subject Matter Experts or "Super Users" for the new system whose knowledge of organizational operations positions them to lead testing, guide configuration decisions, and train their colleagues on new processes. By providing them a path to becoming experts in the new environment rather than obsolete relics of the old one, you channel

their knowledge productively and transform potential resisters into invaluable resources. The Expert who led your accounts payable team's resistance to the new system can become the person who leads your accounts payable team's mastery of it, carrying the same status and respect into the new operational reality.

The Overwhelmed employee is already working 50 hours per week just to keep up with current responsibilities, and the new system doesn't feel like an opportunity for improvement. It feels like one more demand being added to an already overflowing plate during a transition period that will only increase their workload before it eventually decreases it. Their resistance stems not from opposition to the change itself but from legitimate concern about their capacity to absorb learning new systems while maintaining current performance expectations. Handling the Overwhelmed effectively requires showing them the light at the end of the tunnel through concrete demonstrations of how the new system will eventually reduce their burden, using the prototypes from your requirements phase to illustrate specifically how tasks that currently consume hours will take minutes once the transition is complete. Focus on short-term wins that they can experience quickly rather than long-term benefits that feel abstract, and provide tangible support such as temporarily backfilling their position or adjusting performance expectations during the transition period to give them breathing room for learning. The Overwhelmed need to believe that their short-term sacrifice will yield genuine long-term relief, and that belief requires evidence rather than promises.

The Saboteur represents the rarest but most dangerous category: an individual who is actively working to make the project fail, often for personal reasons that may have little to do with the project itself. Perhaps they fear that new systems will make their position redundant, they hold personal grievances against project leaders that make them want to see those leaders fail, or they've determined that their interests are better served by the status quo than by successful transformation. Whatever the underlying motivation, their behavior goes beyond skepticism or overwhelm into deliberate obstruction: spreading misinformation, encouraging others to resist, undermining decisions they've ostensibly agreed to support, or simply failing to complete assigned tasks in ways that create delays they can blame on others. Handling Saboteurs requires direct intervention from leadership rather than the collaborative approaches appropriate for other resistance types. Their manager, supported by your Executive Sponsor, must have a clear, direct conversation that names the specific problematic behaviors, explains why those behaviors are unacceptable, and outlines the consequences that will follow if the behaviors continue. This is not a coaching conversation or a change management intervention. It is a performance management conversation that makes clear the organization will not allow a single individual to hold the success of the entire enterprise hostage to their personal agenda. When confronted directly with consequences, most Saboteurs will either change their behavior or remove themselves from the situation. Either outcome is preferable to allowing sabotage to continue unchecked.

THE REVOLUTION IN ADOPTION: AI-ENABLED TRAINING

Traditional ERP training follows a model that everyone recognizes and almost everyone dreads: herding hundreds of employees into classrooms for days of instruction, walking through generic training materials that may not reflect your organization's specific configuration, and hoping that some fraction of what's presented will stick long enough for employees to apply it when they finally encounter the live system weeks later. Research on learning retention demonstrates why this approach produces such disappointing results: within a week, employees forget approximately 75% of what they learned in traditional classroom training, and by the time they actually need to use that knowledge in their daily work, most of it has evaporated entirely.[85] The gap between training and application is simply too wide for traditional methods to bridge effectively.

AI has been fundamentally transforming this broken model through tools called Digital Adoption Platforms, exemplified by products such as Whatfix and WalkMe, that integrate directly into your new ERP system rather than existing as separate training environments. These platforms function as real-time guidance systems embedded in the actual software employees use, providing contextual help exactly when and where employees need it rather than expecting them to remember classroom instruction from weeks earlier.

Consider how this works in practice when an employee needs to complete a task they haven't performed before or haven't performed recently enough to remember the steps. In the traditional model, they would struggle to recall what they learned in training, perhaps consult a paper manual, or SharePoint document that may or may not reflect current procedures. Potentially, they would call the help desk and wait for assistance, or simply give up and find a workaround that defeats the purpose of the new system. In the AI-enabled model, they click a help icon within the application itself, and the Digital Adoption Platform provides a step-by-step, interactive walkthrough inside the live system, guiding them through each field and each click with contextual explanations of why each step matters until the task is complete. The guidance appears exactly when needed, in the context where it will be applied, at the moment when the employee is motivated to learn because they're trying to accomplish something real.

---

85   Based on the "Forgetting Curve" research by Hermann Ebbinghaus, it is a foundational concept in learning science, which shows a rapid drop-off in memory retention without reinforcement.

The impact of this approach on adoption speed and quality extends across multiple dimensions that directly affect your implementation's return on investment. Organizations deploying Digital Adoption Platforms (DAP) consistently report reducing formal classroom training time by 50% or more because so much learning happens on the job, in the flow of actual work.[86] User proficiency and confidence increase substantially because guidance is available at the moment of need rather than requiring recall of training that occurred weeks earlier. Help desk calls related to basic procedural questions can drop by 50% or more because employees can find answers instantly within the application rather than submitting tickets and waiting for response.[4] New employee onboarding accelerates dramatically because the DAP serves as a personal digital tutor that helps new hires become proficient on complex systems in days rather than the weeks that traditional training requires.

This AI-enabled training approach connects directly to the AI Governance and Readiness Assessment completed during Phase Zero, which should have evaluated your organization's capacity to deploy and support these tools effectively. It also connects to the speed-to-value themes that have run throughout this book: The faster your employees achieve proficiency with new systems, the faster your organization captures the benefits that justified the implementation investment. Every week that employees struggle with unfamiliar interfaces, rely on workarounds rather than intended processes, or

---

86   Forrester Research, "The Total Economic Impact™ of Digital Adoption Platforms," Commissioned Study, 2024.[4] Research from Digital Adoption Platform providers such as Whatfix and WalkMe consistently shows significant reductions in support ticket volume following implementation.

avoid functionality they find confusing is a week when your return on investment remains unrealized.

By investing in an AI-enabled training strategy, you're not merely teaching employees how to use new software. You're also embedding ongoing support into the software itself in ways that ensure employees can use it effectively from their first day on the new system, dramatically compressing the timeline from go-live to genuine operational transformation.

▎KEY TAKEAWAYS

Effective OCM requires moving beyond broadcasting information through official channels to engaging stakeholders through the informal networks where opinions actually form. The following approaches address the reality that employees form their opinions about organizational change through trusted relationships and observable leadership behavior:

- Building a Change Champion network of respected informal leaders who carry accurate information to their peers with credibility that project teams can never achieve
- Ensuring every communication answers the "What's In It For Me?" (WIIFM) question that determines whether messages get absorbed or ignored
- Deploying your Executive Sponsor as a visible, active presence rather than a passive endorser

The school district that discovered their communication efforts had failed despite checking every box learned that effort measured by what you send means nothing; only what gets received, understood, and believed matters for adoption readiness.

Resistance to change emerges from identifiable sources that require different responses. Diagnosing correctly which type of resistance you're facing determines whether your intervention helps or backfires. Skeptics whose concerns are rooted in legitimate experience with past failures need acknowledgment and evidence rather than dismissal. Experts whose identity is threatened by obsolescence of their knowledge need paths to becoming experts in the new environment. Overwhelmed employees need concrete demonstrations that short-term sacrifice yields long-term relief, along with tangible support during the transition period. Saboteurs whose behavior goes beyond resistance into active obstruction require direct leadership intervention with clear consequences. Treating all resistance as the same phenomenon to be overcome through better communication misses the fundamental differences in what's driving resistance and what will resolve it.

AI-enabled training through Digital Adoption Platforms represents a fundamental break from traditional classroom approaches that everyone dreads. Moreover, research has shown they don't work. By embedding contextual guidance directly into the live system, providing help exactly when and where employees need it rather than expecting recall of training from weeks earlier, these tools compress the time to proficiency in ways that directly accelerate your return on investment. The era of herding employees into classrooms, walking through generic materials, and hoping some fraction sticks long enough to matter is ending, replaced by approaches that meet employees where they are and guide them through tasks in the actual flow of work.

▌ YOUR IMMEDIATE ACTION

Within the next 2 weeks, establish the OCM infrastructure that will carry your organization from technical go-live to genuine operational transformation.

Begin by building your Change Champion roster through conversations with managers across the organization who can identify the informal leaders in their areas, the people others turn to when they want to know what's really happening and whether they should support or resist organizational initiatives. These are not necessarily the highest performers or the most senior staff. They're the social connectors whose opinions carry weight with their colleagues. Identify 20–30 of these individuals across functional areas and organizational levels. Next, schedule a special project briefing that makes them the most informed people in the organization and positions them as extensions of your OCM function who will carry accurate information into conversations your project team will never be part of.

The next step is conducting a WIIFM Workshop for each major employee category affected by your implementation. It brings together representatives from each role group to identify the specific ways the new system will improve their daily work experience. For payroll clerks, the benefit might be eliminating manual reconciliation that currently consumes 2 days each period. For procurement officers, it might be mobile approval capabilities that eliminate the backlog that accumulates when they're away from their desks. For field supervisors, it might be real-time visibility into budget status that prevents the surprises that currently disrupt their operations. Transform these specific benefits into simple, powerful talking points that your leaders and Change Champions can use in conversations throughout the organization, ensuring that everyone hears messages tailored to their actual concerns rather than generic project updates that don't connect to their daily reality.

Finally, task your training or HR leadership with scheduling demonstrations from two or three leading Digital Adoption Platform vendors, experiencing firsthand how this technology provides real-time, in-application guidance that transforms the training and support model. Seeing how a DAP walks users through complex tasks with contextual help that appears exactly when needed will fundamentally change how your leadership thinks about training strategy, revealing the gap between traditional approaches that research shows don't work and AI-enabled approaches that meet employees in the flow of their actual work.

The human element of change is where most implementations ultimately succeed or fail, because technical systems that employees don't use effectively or processes that employees resist adopting deliver none of the benefits that justified the investment. By moving beyond outdated communication tactics that confuse sending messages with changing minds, addressing resistance based on accurate diagnosis of its sources, and embracing AI-enabled training that compresses the timeline from go-live to proficiency, you can transform what most projects treat as their greatest risk into your implementation's greatest asset: a workforce that is not merely compliant with new requirements but is genuinely committed to working differently because they under-stand and believe in the benefits that change will deliver.

10

The Independence Principle: Building High-Trust Consulting Relationships

WHAT YOU'LL LEARN

The governance structures you established, the organizational change management strategies you're executing, and the AI-enabled tools you're deploying all depend on a foundational element that determines whether these approaches protect your interests or merely create the illusion of protection, while conflicts of interest undermine your project from within. That foundational element is the independence of the advisors guiding your decisions, a concept that sounds straightforward but proves remarkably difficult to maintain in a consulting ecosystem designed to obscure the relationships that compromise objectivity.

This chapter examines the critical differences between software vendors, system integrators, and truly independent advisors, differences that well-meaning government leaders consistently misunderstand with consequences measured in tens of millions of dollars and years of delay. You'll learn

- Why confusing these roles represents a strategic blunder rather than a semantic quibble
- How to build the high-trust partnerships that enable advisors to deliver the uncomfortable truths you need to hear rather than the reassuring messages that feel better in the moment
- How independence becomes especially critical when adopting AI capabilities that vendors are eager to embed in ways that deepen your dependency
- How to structure advisory contracts in ways that align your partners' financial incentives with your success rather than with project expansion

The relationships you build with external partners will shape every major decision your implementation faces. Ensuring those relationships serve your interests rather than vendor interests is perhaps the most important structural protection you can establish.

THE PHONE CALL THAT REVEALED A FATAL FLAW

In 2023, the CIO of a large eastern US county called me in evident distress about an ERP implementation that had veered dramatically off course despite what she believed were adequate oversight arrangements. The project was running more than a year behind the original schedule and hemorrhaging cash at a rate that had attracted unwelcome attention from the governor's office and legislative oversight committees. "I don't understand what's happening," she said, her voice carrying the particular tension of someone whose professional reputation has become entangled with a failing initiative. "Our independent advisor keeps telling us everything is fine, that we're making progress, that the delays are normal for projects of this complexity. But the project is clearly failing by every objective measure. We're paying them millions for oversight, but they're not protecting us at all."

I asked a simple question that I suspected would reveal the source of her problem: "Who is your independent advisor?"

She named one of the largest, most respected system integrators in the world, a firm whose brand recognition and government experience would seem to make them an obvious choice for oversight responsibilities. What she didn't mention, because she didn't know, was that this same firm maintained "Diamond Level Partner" status with the software vendor whose platform they were supposedly overseeing objectively. It was a relationship that generated tens of millions of dollars in annual revenue for the integrator and created incentives that fundamentally conflicted with the state's interests.

"There's your problem," I told her, wishing the diagnosis were less predictable. "You don't have an independent advisor. You have your vendor's biggest partner providing oversight, a firm whose financial success depends on maintaining that partnership and whose business model requires them to protect the vendor relationship that their entire practice depends on. You've hired the fox to guard the henhouse, given him the keys to the coop, and now you're surprised that feathers are flying everywhere."

This scenario represents the single most common and most costly misunderstanding in government technology, a confusion that I encounter repeatedly despite decades of industry experience that should have made the distinction obvious by now. Well-meaning, intelligent leaders who would never dream of hiring their contractor's business partner to inspect construction quality consistently fail to grasp the fundamental structural differences between the three key players in any implementation ecosystem: the software vendor, the system integrator, and the truly independent advisor. Confusing these roles is not a minor semantic error that practitioners obsess over, while clients reasonably focus on substantive matters. It's a strategic blunder that creates exactly the conflicts of interest leading to budget overruns, missed deadlines, and implementations that consume years of organizational energy without delivering their intended benefits.[87]

The county in my story ultimately had to reset their entire project, terminating the conflicted oversight arrangement and bringing in genuinely independent advisors to assess the damage and chart a path forward. That reset cost them millions in additional expenditure and a 2-year delay beyond their already-extended timeline, consequences that could have been entirely avoided had they understood from the beginning that their independent advisor was structurally incapable of providing the objective oversight they were paying for.

87　This scenario is a composite based on multiple real-world examples documented in state auditor reports and news articles where conflicts of interest in IT procurement led to project resets and significant financial losses.

DECONSTRUCTING THE ECOSYSTEM: UNDERSTANDING WHAT DRIVES BEHAVIOR

Protecting yourself from the scenario that trapped that state CIO requires understanding the business model of each player in the implementation ecosystem, because financial incentives dictate behavior far more reliably than professional ethics or good intentions. People and organizations consistently act in accordance with how they get paid, and understanding those payment structures reveals whose interests get served when conflicts arise.

Consider an analogy that makes these distinctions tangible: Imagine you're building a custom home, a complex project involving multiple parties with different roles and different relationships to your success.

The software vendor in this analogy resembles the company that manufactures the lumber, pipes, wiring, and other materials your home requires. Their primary goal is to sell you as many of their high-quality materials as possible, ideally including premium product lines and optional additions that increase the total sale. They want your project to succeed in the sense that successful projects generate positive references and repeat business; however, their fundamental loyalty runs to their shareholders and their product line rather than to your specific project outcomes. They are not your project partner in any meaningful sense but a supplier of key components whose interests align with yours only to the extent that your success reflects well on their products. When their revenue interests conflict with your project interests, expecting them to prioritize your interests misunderstands the nature of the relationship.

The system integrator occupies a more complex position in this ecosystem, one that creates the confusion that trapped the CIO. In the home-building analogy, the SI resembles a large construction company that maintains exclusive partnership agreements with the materials manufacturer and building its entire business around expertise in constructing homes using those specific materials. Their business model centers on selling you the billable hours of their carpenters, plumbers, electricians, and project managers, which means their revenue increases when projects grow larger and take longer. They operate under dual loyalty that creates structural tension: They are genuinely loyal to you as the paying client whose satisfaction they need for references and repeat business, but they are also fundamentally loyal to the manufacturer whose partnership their entire practice depends on and whose products they cannot objectively evaluate because doing so would threaten the relationship that makes their business viable.

A system integrator cannot serve as your independent advisor because their financial success is directly tied to both the vendor's success and the size of your project. More complexity doesn't represent a problem for an SI—it represents a profit center that generates additional billable hours. Change orders don't signal failures in planning or execution—they signal opportunities for revenue expansion. Timeline extensions don't threaten the SI's success—they extend the engagement and increase total contract value.

These aren't accusations of bad faith or unethical behavior. They are simply the mathematical reality of how the SI business model works. When you ask an SI to provide independent oversight, you're asking them to potentially reduce their own revenue by identifying unnecessary complexity, flag problems with their vendor partner that could damage a relationship worth millions in annual revenue, and accelerate timelines that would reduce their billable hours. Expecting them to act against these powerful financial incentives based on professional ethics alone misunderstands how organizational behavior works.

The truly independent advisor occupies a fundamentally different position in this ecosystem, one that aligns their interests with yours rather than with vendors or project expansion. In the home-building analogy, this is the master architect and general contractor you hire before engaging anyone else, one who maintains no exclusive relationships with materials manufacturers or construction firms, and whose only product is expert, unbiased advice about how to achieve your objectives. Their loyalty runs entirely to you as the client because their business doesn't depend on any vendor relationship, and their revenue doesn't increase when your project grows larger or takes longer. Their sole incentive is ensuring your project succeeds according to your definition of success, on time and on budget, because their reputation depends entirely on client outcomes rather than on maintaining vendor partnerships or maximizing billable hours.

Research on public sector technology projects, including studies by the Brookings Institution and the Government Accountability Office, consistently identifies the presence of a truly independent third-party quality assurance function as one of the strongest predictors of project success.[88] This finding makes intuitive sense once you understand the incentive structures: The independent advisor is the only party in the ecosystem whose financial interests are perfectly aligned with

---

88   The Brookings Institution, "Driving Public Sector Success Through Outcome-Based Contracting," June 2022.

yours and the only party who profits when you succeed quickly and efficiently rather than when projects expand and extend.

FOSTERING PARTNERSHIPS BASED ON TRUST

Once you've engaged a truly independent partner whose structural position aligns their interests with yours, building the high-trust relationship that enables effective oversight becomes the essential next step. This relationship differs fundamentally from typical client–vendor transactions, where defined deliverables are exchanged for defined payments with limited ongoing engagement. An effective advisory partnership involves navigating years of complex decisions together, many of which will be contentious, uncomfortable, or politically fraught. That navigation requires trust deep enough to sustain difficult conversations and honest enough to surface problems before they become crises.

This trust is built on a foundation of radical transparency flowing in both directions. Your independent advisor must have a seat at every table where significant decisions are made and access to all information about project status, organizational dynamics, and emerging challenges, whether that information reflects well on the project or reveals problems that stakeholders would prefer to minimize. They must be empowered to deliver unwelcome news to the Executive Sponsor without fear that honest reporting will cost them the engagement, and they must be confident that their assessment of project reality will be heard even when it contradicts what other parties are reporting. In turn, you must be transparent with your advisor about the internal politics shaping decisions, the budget constraints limiting options, and the organizational fears influencing how stakeholders respond to change. An advisor working with incomplete or sanitized information cannot protect you effectively, no matter how capable they might be, because the problems they don't know about are the problems they cannot help you address.

This transparency naturally leads to what effective partnerships embrace rather than avoid—constructive conflict about decisions, priorities, and interpretations of project status. A good advisory partner doesn't simply agree with your assessments, validate your preferences, and tell you what you want to hear. They challenge assumptions that may not withstand scrutiny; push back when department heads make demands that would compromise project integrity; and tell you hard truths about your team's readiness, your timeline's realism, or your organization's capacity to absorb change. This conflict is constructive precisely because it surfaces problems while they remain manageable and forces reconsideration of decisions that might otherwise proceed unchallenged toward predictable failure.

As a leader, your responsibility is to create an environment where your advisor is rewarded for surfacing problems and challenging comfortable assumptions rather than punished for delivering messages that create discomfort. When your advisor tells you something you don't want to hear, you are receiving exactly what you paid for: the unvarnished truth that internal staff may be reluctant to deliver and that vendor partners are structurally incapable of providing. If you respond to uncomfortable truths by questioning your advisor's judgment, minimizing their concerns, or creating consequences that discourage future candor, you systematically destroy the value that independence was supposed to provide. The advisor learns that honesty creates problems while reassurance creates comfort, and rational actors respond to those incentives by telling you what you want to hear rather than what you need to know.

The advisory relationships that create the most value are those where clients genuinely welcome challenge, where advisors feel safe delivering difficult messages, and where both parties understand that temporary discomfort from honest assessment prevents permanent damage from problems that compounded unchecked. Building this kind of relationship requires conscious effort from both parties, but the protection it provides justifies that effort many times over.

THE AI CO-INNOVATION TRAP

Nowhere do the dynamics of independence become more critical than in the adoption of artificial intelligence, where every vendor and system integrator now brands themselves as an "AI partner," eager to help you co-innovate and transform your organization through cutting-edge capabilities. This positioning sounds appealing, and the AI capabilities these partners offer may be genuinely impressive, but the structure of such partnerships creates dependency traps that will constrain your organization for years after the initial enthusiasm fades.

When you co-innovate with a vendor or their system integrator partner, you are almost always building solutions that become inextricably tied to their proprietary platform in ways that may not be apparent until you attempt to change direction. They will help you build AI tools that are genuinely useful, solve real problems, and demonstrate the transformative potential of AI in government operations. But those tools will only work within their software ecosystem, requiring their platform to function and their ongoing involvement to maintain. Every successful AI implementation you build with them deepens your dependency on their platform, increases the cost of eventually switching to alternatives, and ensures that their revenue stream continues regardless of whether their ongoing value justifies their ongoing fees.

The intellectual property dynamics of vendor-led AI development create additional concerns that organizations often discover too late. When you co-develop AI capabilities with a vendor, the resulting innovations frequently become vendor intellectual property that they can sell to other clients, meaning you funded development of capabilities that your competitors will access without bearing any of the development cost. The data you provide for training AI models may become entangled with vendor systems in ways that complicate future data portability. The expertise your staff develops becomes expertise in vendor-specific tools rather than transferable AI capabilities that would remain valuable if you eventually change platforms.

A truly independent advisor helps you build an AI strategy that remains platform-agnostic, one that leverages AI capabilities without creating the dependency that vendor partners are structurally incentivized to deepen. They help you identify business problems that AI could genuinely solve rather than simply finding applications for whatever AI features your current vendor happens to be promoting. They can introduce you to a wide range of AI tools, including open source options and innovative smaller vendors whose solutions might better fit your specific needs, rather than limiting your consideration to the embedded AI your ERP provider wants you to adopt. They help you structure AI pilot projects in ways that prove value and build organizational capability without creating long-term dependency, ensuring that the data you generate and the AI models you develop remain your intellectual property rather than becoming entangled with vendor systems.

Your goal in AI adoption should be using these powerful capabilities to become more self-sufficient, more capable of solving your own problems, and more flexible in how you deploy technology to serve your mission. Vendor-led AI development pushes in exactly the opposite direction—toward deeper dependency, reduced flexibility, and ongoing reliance on external parties for capabilities that could be internalized. An independent advisor is the only partner in the ecosystem whose incentives align with your goal of self-sufficiency rather than the vendor goal of expanded dependency.

STRUCTURING CONTRACTS THAT MAINTAIN ALIGNMENT

A skeptical reader might raise some fair questions at this point: If you hire an independent advisor, aren't you simply adding another consultant whose business model involves billing you for hours and whose incentives might therefore favor extended engagements and expanded scope? Who watches the watchers, and how do you ensure that your independent partner accelerates your project rather than becoming another cost center?

This skepticism is healthy because it recognizes that independence alone doesn't guarantee aligned incentives. The structure of the engagement determines whether independence translates into value or merely substitutes one set of misaligned incentives for another. If you structure your independent advisor's contract identically to how you structure system integrator contracts, with time-and-materials billing that rewards hours worked rather than outcomes achieved, you risk recreating the exact misalignment you're trying to avoid.

Ensuring your independent partner remains an accelerator whose interests align with your success requires structuring engagements in ways fundamentally different from typical consulting contracts.

Oversight functions should be structured as fixed-fee arrangements rather than time-and-materials billing because the work involved is relatively steady-state. A fixed monthly fee for PMO and oversight responsibilities removes the incentive for advisors to find work that generates billable hours without adding value, creating incentives instead to keep the project healthy and moving efficiently because their compensation doesn't increase when problems arise. They get paid to maintain project health rather than to generate the documentation and meetings that time-based billing rewards regardless of whether that activity contributes to project success.

A portion of advisory compensation should be tied to specific project outcomes rather than merely to attendance at meetings and production of deliverables that may or may not contribute to success. Consider structuring contracts with milestone payments tied to questions such as

- Whether the requirements phase completed on schedule with documented stakeholder approval
- Whether the vendor selection process concluded without protests that signal process failures
- Whether testing phases completed with zero Severity 1 defects remaining unresolved

- Whether go-live occurred within defined tolerance of the original timeline

When advisors profit more from successful outcomes than from extended engagements, their incentives align with your calendar rather than conflicting with it.

Effective advisory contracts should include what practitioners call a "sunset clause," which explicitly plans for the relationship's conclusion from its beginning. A good independent advisor should be working toward their own obsolescence from day one, building your internal capability to eventually perform oversight functions yourselves rather than creating permanent dependency on their involvement. The contract should include a formal knowledge transfer phase where advisors train your internal staff to operate a Center of Excellence post-go-live, with success criteria that define what "ready to operate independently" really means. If an advisor is still running your daily operations 3 years after go-live, something has gone wrong with either the knowledge transfer process or the advisor's willingness to complete their mission and move on.

By structuring contracts using these elements, you align the advisor's financial interests with your project timeline and your organizational development goals. They profit most when you succeed on schedule and develop the internal capability to sustain that success without ongoing external dependency.

▌KEY TAKEAWAYS

- Understanding business models reveals whose interests get served when conflicts arise, because organizations consistently behave in accordance with how they generate revenue regardless of professional ethics or stated intentions. A software vendor's business model centers on selling products, which means their interests align with larger implementations using more of their capabilities rather than with right-sized implementations that meet your actual needs.

- A system integrator's business model centers on selling billable hours to implement vendor products, which means their interests align with larger, longer, more complex projects rather than with efficient implementations that minimize their involvement. A truly independent advisor's business model centers on reputation for client success, which means their interests align with your success rather than with project expansion. Confusing these fundamentally different business models and expecting vendor partners to provide the objective oversight that only independence enables represents the most expensive mistake government technology leaders consistently make.

- High-trust advisory partnerships are built on radical transparency and constructive conflict rather than comfortable agreement and reassuring messages. Your advisor needs access to all information about your project and organization, including the uncomfortable truths that stakeholders prefer to minimize, because they cannot protect you from problems they don't know about.

- You need an advisor willing to challenge your assumptions, push back on decisions that may not withstand scrutiny, and deliver hard truths about readiness and risk that internal staff may be reluctant to voice. When your advisor tells you something you don't want to hear, you're receiving exactly what their independence was supposed to provide, and responding in ways that discourage future candor systematically destroys the value you're paying for.

- Independence becomes especially critical in AI adoption, where every vendor positions themselves as a transformation partner eager to help you co-innovate in ways that inevitably deepen your platform dependency. Vendor-led AI development builds capabilities that only work within vendor ecosystems, generates intellectual property that may belong to the vendor rather than to you, and creates integration patterns that make future platform changes increasingly difficult and expensive.

- An independent advisor helps you build AI strategies that remain platform-agnostic, leverage diverse tools including open-source options, and develop internal capabilities supporting long-term self-sufficiency rather than deepening the dependency that vendor partners are structurally incentivized to create.

▌ YOUR IMMEDIATE ACTION

Within the next 2 weeks, assess whether your current advisory relationships provide the independence you need or merely create the appearance of objective oversight while conflicts of interest compromise the guidance you receive.

Begin by conducting a vendor relationship audit of every external consulting firm involved in your major technology initiatives, investigating the official partnership status each firm maintains with your key software vendors. Most consulting firms publicly announce their partnership levels with major vendors, and a few hours of research will reveal whether your advisors hold Platinum, Diamond, or other elite partnership designations with the vendors whose products they're supposedly helping you evaluate objectively.

If your oversight partner maintains partnership status that generates significant revenue from the vendor they're overseeing, you have identified a structural conflict of interest. No amount of professional ethics can fully resolve it, and such conflict of interest requires either renegotiating the relationship or engaging truly independent oversight.

Next, assess the quality of your most critical consulting relationships by asking yourself when your advisor last brought you genuinely bad news about project status, organizational readiness, or decisions that needed reconsideration. Consider when they last challenged one of your assumptions in ways that made you uncomfortable or pushed back on a direction you preferred in ways that required you to defend your thinking.

If you struggle to identify recent examples of constructive conflict, you may not have a high-trust partnership that delivers the honest assessment independence is supposed to provide. Instead, you may have a high-cost echo chamber that tells you what you want to hear while problems compound unaddressed.

Finally, work with your procurement and legal teams to draft standard contract language for all future advisory engagements that requires firms to disclose any financial relationships with software vendors and system integrators before engagement begins, that structures compensation around outcomes rather than hours, and that includes knowledge transfer requirements ensuring advisory relationships have defined endpoints rather than becoming permanent dependencies. Making independence a nonnegotiable criterion for oversight work, verified through disclosed relationships and maintained through aligned incentive structures, establishes the foundation that enables everything else in your implementation to function as intended.

Building a relationship with a truly independent partner whose incentives align with your success represents the most important structural investment you can make in your modernization journey. Independence is the foundation that makes honest assessment possible, enables constructive conflict to surface problems before they become crises, and ensures every decision from vendor selection through AI strategy is made with only one interest in mind—yours.

11

AI Comes of Age:
The New Rules of Government Technology

WHAT YOU'LL LEARN

The independence principles we've explored throughout this book take on heightened importance as artificial intelligence transforms what's possible in government technology. The governance structures from Chapter 8, the organizational change strategies from Chapter 9, and the advisory relationships from Chapter 10 all become more critical when AI capabilities enter the equation, because AI simultaneously accelerates outcomes and multiplies hidden risks in ways that demand the kind of neutral oversight only independence provides.

This chapter examines what distinguishes AI projects that deliver genuine value from the 80% that stall or fail—a failure rate that RAND analysts have documented as twice that of ordinary IT initiatives.[89] You'll learn:

---

89   RAND Corporation, "Root Causes of Failure for Artificial Intelligence Projects and How They Can Succeed," Research Report RRA2680-1, 2024.

- Why the traditional rip-and-replace model for government technology is giving way to composable modernization approaches that can extend legacy systems with targeted AI capabilities rather than requiring wholesale replacement How federal mandates have established governance frameworks that cascade through all levels of government, creating expectations you should meet regardless of direct legal requirements.

- Why AI's lower barrier to entry paradoxically raises the bar for due diligence, making independent oversight more essential rather than less as these powerful tools become accessible to anyone with a subscription.

THE PARADOX OF AI IN GOVERNMENT: ADOPTION SURGE, FAILURE EPIDEMIC

By mid-2025, almost two-thirds of U.S. states and nearly half of large local agencies are actively piloting or procuring AI-enabled extensions for their finance, human resources, and permitting systems.[90] The technology has moved from keynote presentations and innovation labs into the daily operations of government agencies seeking to do more with constrained resources.

AI-powered chatbots now handle citizen inquiries around the clock without requiring staff to work nights or weekends. Predictive analytics identify infrastructure maintenance needs before failures

---

90 Unanet, "AI Year in Review: Advancements and Predictions for 2025," December 2024.

occur, preventing the emergency repairs that consume enormous budgets and disrupt public services. Automated document processing reduces permit approval timelines from weeks to days while improving accuracy and compliance with regulatory requirements.

Yet this adoption surge coexists with an epidemic of AI project failure that should give every government leader pause. RAND's analysis of AI initiatives across sectors found that more than 80% stall before delivering intended value, making AI projects twice as likely to fail as traditional IT implementations.[91] A 2024 survey of federal agencies found that chatbot pilots had the lowest progression rate to production of any AI category, with only 57% advancing beyond pilot stage because agencies lacked frameworks to certify privacy and bias controls.[92] The Stanford HAI AI Index 2025 documented that while AI investment continues breaking records, the gap between pilot deployments and scaled production systems remains stubbornly wide.[93]

These seemingly contradictory trends resolve into a coherent pattern once you understand what distinguishes successful AI initiatives from expensive experiments. Analysis of government AI adoption reveals a stark difference: Agencies working with independent consultants achieve 85%–90% user adoption rates for AI-powered systems, while vendor-led implementations struggle to reach 60%–70% adoption.[94]

---

91   RAND Corporation, RRA2680-1.

92   General Dynamics Information Technology Survey, "Federal AI Implementation Progress," 2024.

93   Stanford University Human-Centered Artificial Intelligence, "AI Index Report 2025."

94   Avèro Advisors analysis of government AI adoption outcomes, 2025.

The difference isn't technical capability—vendors often bring sophisticated AI tools and skilled technologists. The difference is strategic alignment and stakeholder engagement that independent consultants prioritize because their success depends on client outcomes rather than software sales or billable hours.

The RAND research identified the strongest predictors of AI project success: clear problem framing that starts with operational needs rather than technology capabilities, and independent governance that provides objective assessment of progress and risks.[95] These findings echo everything this book has emphasized about the importance of Phase Zero preparation, vendor-neutral evaluation, and governance structures that specifically serve your interests. AI doesn't change these fundamentals but amplifies their importance because both the potential benefits and the potential harms are larger.

FROM MONOLITH TO MOSAIC: THE COMPOSABLE ERP REVOLUTION

For three decades, government ERP projects resembled interstate highway construction: enormous budgets, disruptive multiyear timelines, and little visible benefit until the end when everything finally came together or collapsed into expensive failure. That model is collapsing under converging pressures that make incremental, targeted approaches both possible and necessary.

---

95   RAND Corporation, RRA2680-1.

Legacy systems now attract escalating cyber risk and audit findings, while consuming up to 80% of many IT operating budgets on maintenance that delivers no new capabilities.[96] Fiscal headwinds are tightening as state and local governments face budget constraints that make hundred-million-dollar replacement projects increasingly difficult to fund. And constituent expectations have shifted from acceptance of government's technology limitations to demands for the same responsive, intuitive digital experiences they receive from commercial services.

Instead of massive overhauls that take years to deliver, agencies are increasingly stitching together what Gartner calls "composable ERP": a mosaic of interoperable services that can be swapped or upgraded without disturbing the core transactional systems that must remain stable.[97] A Colorado utility district recently overlaid a rate-modeling service onto its 1990s billing platform. The extension went live in 7 weeks, cost less than 1% of the estimate for a full replacement, and delivered real-time simulation capabilities for commissioners evaluating rate structure alternatives. This approach represents a fundamental shift from the binary choice between maintaining aging legacy systems with increasing costs and capability gaps or undertaking high-risk, high-cost replacement projects.

The technology advances enabling this shift include AI-enabled microservices that provide specific functionality through modular, interoperable components; micro-SaaS solutions that address par-

---

96   National Conference of State Legislatures, "Artificial Intelligence in Government: The Federal and State Landscape," 2024.

97   Gartner, Inc., "Composable ERP Strategy for Government," 2024; Public Sector Network, "Modernize Government ERP Systems to Improve Innovation and Agility," 2025.

ticular business needs without requiring platform replacement; and low-code and no-code platforms that enable rapid development of custom capabilities without extensive programming. In addition, API management frameworks facilitate integration across disparate systems, and data virtualization tools provide unified access to information regardless of where it physically resides. Together, these technologies enable agencies to maintain core legacy systems that continue functioning reliably while strategically extending them with targeted modern capabilities addressing specific operational needs.

This composable approach changes the modernization calculus in ways that favor independence and reduce vendor leverage. When you can extend existing systems incrementally rather than replacing them wholesale, you're not locked into multiyear commitments with single vendors who control your technology future. When you can deploy targeted solutions in weeks rather than years, you can experiment, learn, and iterate rather than betting everything on specifications defined before you really understood what you needed.

When you can swap components that aren't working without disrupting everything else, vendors must continue earning your business rather than coasting on switching costs that trap you in relationships regardless of performance.

WHERE AI ACTUALLY DELIVERS VALUE IN GOVERNMENT OPERATIONS

The headlines about AI often dwell on science-fiction scenarios and speculative transformations, but actual deployments in government cluster around four practical patterns that deliver measurable value without requiring revolutionary change to established operations.

Low-code orchestration tiers that broker data between legacy ERP systems and new AI models represent the most common pattern.[98] Platforms such as Microsoft Power Platform, Appian, or ServiceNow allow finance analysts and operations staff rather than programmers to adjust workflows when regulations change or new requirements emerge. These platforms provide the integration layer that connects stable transactional systems to AI capabilities without requiring modifications to core applications that must remain reliable.

Anomaly and pattern detection using machine learning to flag procurement exceptions, payroll outliers, or budget variances represents the second major deployment pattern. These capabilities don't replace human judgment. They focus human attention on the transactions that most warrant review rather than requiring staff to manually examine every transaction looking for problems. A utility billing system might process millions of routine bills automatically, while flagging the small percentage with unusual consumption patterns or payment anomalies for staff review. A procurement system might identify vendors whose pricing deviates from historical patterns or whose invoices contain characteristics associated with fraud in prior investigations.

Generative assistants for drafting routine documents, answering internal policy questions, or summarizing large meeting transcripts represent the third pattern, one that has expanded rapidly with the maturation of large language model technology. These capabilities

---

98   DLT Solutions, "ERP Modernization for Government," White Paper, April 2025.

reduce the time staff spend on routine writing and research tasks without replacing the human judgment required for consequential decisions. A permit application system might draft initial response letters that staff review and modify rather than writing from scratch. A human resources system might answer common employee questions about benefits or policies without requiring staff to respond individually to repetitive inquiries.

Citizen-facing services like chatbots, appointment scheduling, and automated status updates represent the fourth pattern, extending government's accessibility beyond traditional business hours without requiring overnight staffing. These deployments work best when they handle genuinely routine interactions while maintaining clear pathways for escalation to human staff when situations exceed automated capabilities. The agencies achieving the best results with citizen-facing AI have invested heavily in understanding the boundary between interactions that automation can handle effectively and those that require human engagement, designing their systems around that boundary rather than pushing automation into territory where it creates frustration.

What distinguishes successful deployments across all four patterns is a characteristic that connects directly to this book's central themes: They've partnered with advisors who understand that AI success depends more on human change management than algorithmic sophistication. Vendor teams focus on software installation and technical configuration because that's what vendors sell. Independent consultants focus on organizational transformation that enables sustainable value creation because that's what clients need to achieve their objectives.

THE REGULATORY LANDSCAPE: FRAMEWORKS THAT APPLY WHETHER REQUIRED OR NOT

The federal AI governance framework has evolved substantially since the Biden administration issued OMB Memorandum M-24-10 in March 2024, establishing requirements for federal agencies to designate Chief AI Officers, inventory AI use cases, and implement minimum risk management practices for systems affecting public rights and safety.[99] In April 2025, the Trump administration replaced that guidance with OMB Memoranda M-25-21 and M-25-22, maintaining core governance structures, while emphasizing accelerated adoption, reduced bureaucratic burden, and preference for American-made AI solutions.[100]

The bipartisan continuity on core governance elements signals emerging consensus rather than partisan preferences likely to reverse

---

99　U.S. Office of Management and Budget, Memorandum M-24-10, "Advancing Governance, Innovation, and Risk Management for Agency Use of Artificial Intelligence," March 28, 2024.

100　U.S. Office of Management and Budget, Memoranda M-25-21 and M-25-22, April 3, 2025.

with future administrations. Both frameworks require designated AI leadership with authority to coordinate adoption across the organization. They mandate inventories of AI use cases that enable oversight and accountability. They also establish risk management practices proportional to potential impact, with heightened requirements for what M-25-21 calls "high-impact AI" that serves as a principal basis for decisions significantly affecting rights or safety. And both emphasize procurement approaches that prevent vendor lock-in and protect government data and intellectual property rights.

For state and local government leaders, what matters is not the specific federal compliance requirements, which don't apply directly to your operations, but the framework these memoranda establish for responsible AI adoption. Current federal procurement guidance applying to solicitations after October 2025 requires contractual provisions around data ownership, vendor lock-in prevention, and ongoing performance monitoring that will likely cascade into standard expectations for government technology procurement broadly.[101] States have forged ahead with their own frameworks, with 45 state legislatures introducing more than 550 AI-related bills in 2025 addressing everything from bias testing requirements to disclosure obligations when AI influences consequential decisions.[102]

Agencies that wait for legal requirements before implementing governance frameworks will find themselves scrambling to comply retroactively with mandates that are easier to build in from the beginning. Those that establish appropriate governance now, whether legally required or not, will find procurement easier, oversight simpler,

---

101 OMB Memorandum M-25-22, Section 4.

102 Stanford HAI AI Index 2025; National Conference of State Legislatures AI legislation tracking.

and outcomes better because the disciplines these frameworks impose genuinely contribute to AI project success regardless of their regulatory status.

THE WORKFORCE EQUATION: SKILLS, ANXIETY, AND THE PATH TO ADOPTION

Surveys of public-sector employees reveal enthusiasm and anxiety in equal measure about AI's arrival in government operations. Forty-seven percent feel their agency is behind on AI adoption relative to peer organizations and private sector comparators. Yet 60% cite skills gaps as the main obstacle to moving faster, expressing concern that neither they nor their colleagues possess the knowledge required to work effectively with AI-enhanced systems.[103]

These data underscore a truth that technology enthusiasts often overlook: AI's promise hinges on people, not code. Implementation plans that dedicate at least 15% of project budgets to change leadership, training, and stakeholder engagement routinely achieve 68% higher user adoption rates than those that treat human factors as afterthoughts.[104] The pattern holds across deployment types and organizational contexts because technology that people don't use or don't use effectively delivers none of its intended benefits regardless of how sophisticated its algorithms might be.

Successful agencies deliver what practitioners call role-based learning journeys that help each employee understand how AI affects

---

103 Boston Consulting Group, "Benefits of AI in Government," 2025.

104 Project Management Institute, "Why Most AI Projects Fail," 2024.

their specific work rather than providing generic training about AI concepts. A payroll clerk practices approving AI-flagged exceptions inside a safe sandbox environment where mistakes don't affect real transactions, building confidence through hands-on experience before encountering the live system. A union steward sees how transparent audit logs document every AI-influenced decision, understanding how the technology can be monitored for the bias they reasonably worry about. Council members preview dashboards that convert AI model outputs into service-level metrics they can use for oversight, understanding how to ask questions about AI performance without needing technical expertise.

When that narrative is missing—when employees don't understand how AI will affect them or can't see themselves succeeding in the AI-enhanced environment—projects unravel even if the software technically performs as designed. The Marin County SAP implementation demonstrated how inadequate change management can doom technically sound deployments. The system worked as configured but users rejected it because nobody helped them understand why the changes were happening or how to succeed in the new environment.[105]

AI also creates new categories of stakeholder concern that traditional change management approaches don't address. Employees worry about job displacement as AI automates tasks they currently perform. Union representatives question whether AI monitoring violates privacy rights or enables unfair performance assessment. Citizens groups raise concerns about algorithmic bias in government decision-making that affects housing applications, benefit eligibility,

---

105 Panorama Consulting, "Lessons Learned from Government ERP Failure," case study analysis.

or enforcement priorities. These concerns require change management strategies that go beyond traditional software training to address fundamental questions about how AI should and shouldn't be used in government operations.

Independent consultants help agencies develop what practitioners call the human-AI partnership model, where technology augments human judgment rather than replacing it in consequential decisions. An AI system might flag potential compliance violations in procurement transactions, analyzing patterns across thousands of purchases far faster than any human could, but human reviewers make final determinations about whether violations occurred and what corrective actions are appropriate.

This approach maintains human accountability while leveraging AI efficiency, addressing both operational needs for faster processing and political concerns about automated government decision-making. The agencies that struggle are those that let vendors drive toward full automation without the governance structures to maintain appropriate human oversight where it matters.

INDEPENDENT ADVISORS AS AI NAVIGATORS

AI's lower barrier to entry paradoxically raises the bar for due diligence. Anyone can subscribe to a GPT-powered service or deploy a low-code AI application, but the probability of silent bias, data exfiltration, or vendor lock-in rises exponentially when procurement shortcuts the rigor that AI's power demands.[106] Independent advisors bridge this gap in ways that vendors structurally cannot provide because vendor interests don't align with the scrutiny that effective AI governance requires.

Independent advisors interpret the evolving statutory landscape and translate requirements into contract language that vendors must honor. Illinois's bias testing law, OMB's data rights clauses, local open-records rules that apply to AI-generated documents, state requirements for human review of high-impact decisions—vendors may acknowledge these requirements exist but have little incentive to make compliance easy or to flag potential violations. Advisors whose success depends on client outcomes rather than vendor relationships maintain focus on compliance that procurement officers may not have expertise to enforce independently.

Independent advisors maintain multivendor evaluation capabilities that prevent what practitioners call "demo theatre." Vendors optimize demonstrations for impressive visual impact rather than actual operational fitness. Because advisors work across multiple vendors and platforms, they can benchmark claims against observed performance in comparable deployments, identify red flags that procurement officers seeing a vendor for the first time might miss, and structure evaluations around realistic scenarios that expose limitations vendors prefer to gloss over.

Independent advisors integrate AI checkpoints into the same PMO cadence that already governs schedule and budget, ensuring that ethics review and bias assessment become standing agenda items rather than side memos that get lost in implementation urgency. They track what some practitioners call AI ethics heatmaps alongside cost

---

106 Relyance AI, "AI Governance Examples," 2024.

variance, monitoring where AI decision-making is concentrated and whether appropriate human oversight is being maintained.[107]

Independent advisors document decision lineage that gives executives defensible narratives when audits or council hearings probe controversial outcomes. When an AI-influenced decision generates public concern, the question will be whether government officials exercised appropriate oversight and followed established processes. Documentation created by interested parties such as vendors is inherently suspect. Documentation from independent advisors whose role included monitoring carries different weight.

Furthermore, independent advisors fortify human capital plans, upskilling staff into algorithm supervisors rather than technology casualties, thereby sustaining adoption long after consultants leave. The goal isn't permanent dependency on external advisors but building internal capability that enables the organization to manage AI effectively on its own. McKinsey's research on AI adoption confirms that organizations investing in workforce transformation alongside technology deployment achieve substantially better outcomes than those treating technology and people as separate workstreams.[108]

The payoff from independent AI oversight is measurable. Across 25 states surveyed by NASCIO, programs with independent PMOs spent 1%–3% of project budgets on neutral oversight yet realized lifecycle savings averaging 12%.[109] RAND's analysis corroborates that clear problem framing and independent governance are the strongest

---

107 Third Stage Consulting, "AI Strategy Governance," 2024.

108 McKinsey & Company, "Superagency in the Workplace: Empowering People to Unlock AI›s Full Potential at Work," 2025.

109 NASCIO survey data on independent PMO effectiveness, 2024.

predictors of AI project success. The investment in independence pays for itself many times over through failures avoided, risks surfaced before they metastasize, and outcomes achieved rather than promised.

GEORGIA'S REGISTRY MODEL: TRANSPARENCY THAT ACCELERATES ADOPTION

While many agencies struggle to move AI initiatives from pilot to production, Georgia has demonstrated how governance frameworks can actually accelerate rather than impede adoption. The state's AI Registry publishes every model's purpose, data source, and risk tier in a format accessible to legislators, advocacy groups, and citizens.[110] The approach invites civil society review rather than treating AI deployment as a technical matter outside public scrutiny.

The results contradict the assumption that transparency and governance slow innovation. By establishing clear processes for reviewing and approving AI deployments, Georgia reduced the legislative pushback that had stalled previous initiatives, with AI-related inquiries from legislators dropping 30% once the registry provided answers to questions that had previously required individualized briefings. Budget approval for subsequent AI projects accelerated because appropriators could see how the state was managing risks rather than worrying about unknown exposure from opaque technology deployments.

---

110 Georgia Technology Authority AI Registry; Government Technology coverage, 2024.

The registry model demonstrates that governance isn't an obstacle to AI adoption but a foundation that makes adoption sustainable. Organizations that try to rush AI into production without governance frameworks frequently find themselves retreating when problems emerge, having to pause or roll back deployments, while they build the oversight mechanisms they should have established from the beginning. Organizations that establish governance first can move confidently because they have mechanisms to identify and address problems before they become crises.

Colorado has implemented a similar approach, with an AI Council that mandates board approval for every high-impact deployment while maintaining an open registry that documents what AI systems are in use and how they're being governed.[111] The approach creates accountability that elected officials can point to when constituents raise concerns about algorithmic decision-making, while giving agency staff clear processes for bringing AI initiatives forward rather than ad hoc approval paths that vary depending on who happens to be paying attention.

▌ KEY TAKEAWAYS

- AI projects fail at twice the rate of ordinary IT initiatives not because the technology is immature but because organizations deploy it without the governance frameworks that success requires. The strongest predictors of AI project success are clear problem framing that starts with operational needs rather than technology capabili-

---

111 Colorado AI Council governance framework documentation.

ties, and independent governance that provides objective assessment throughout deployment.

- Organizations working with independent consultants achieve dramatically higher adoption rates than those relying on vendor-led implementation because independent advisors focus on organizational transformation rather than software configuration. They focus on the human factors that determine whether technically capable systems actually deliver value.

- The composable ERP revolution enabled by low-code platforms, microservices, and API frameworks creates alternatives to the traditional replace-or-maintain binary that has driven government technology decisions for decades. Agencies can now extend legacy systems with targeted AI capabilities that address specific operational needs, deploying solutions in weeks rather than years at a fraction of traditional costs.

- This approach reduces vendor leverage by eliminating the all-or-nothing dynamics that gave vendors power over organizations locked into multiyear implementations with limited alternatives. It also aligns naturally with incremental governance approaches where organizations can learn from small deployments before scaling, rather than betting everything on upfront specifications.

- Federal AI governance frameworks have established expectations that cascade through all levels of government regardless of direct legal applicability. The core elements of designated AI leadership, use case inventories, risk management proportionate to impact, and procurement provisions preventing vendor lock-in represent emerging consensus that organizations should implement whether currently required or not. These frameworks contribute to AI project success independent of their regulatory status because the disciplines they impose address the failure modes that doom AI initiatives across sectors and contexts.

▎YOUR IMMEDIATE ACTION

Within the next 30 days, inventory every AI capability currently touching your operations, including scripts, bots, macros, and third-party services that incorporate AI components. For each capability, document its purpose, the data it accesses, the decisions it influences, whether human oversight mechanisms exist, and who is accountable for its performance. This inventory creates the foundation for governance regardless the framework your organization eventually adopts. The discovery process frequently reveals AI usage that leadership wasn't aware of and hadn't evaluated for appropriateness.

Establish an AI governance checkpoint within your existing PMO structure rather than creating separate oversight mechanisms that fragment accountability. The PMO already tracks schedule, budget, and risk for major initiatives. Adding AI-specific checkpoints to that cadence ensures that algorithmic ethics and bias assessment receive the same attention as traditional project concerns. Define criteria for what constitutes high-impact AI requiring enhanced review versus routine applications that can proceed through standard approval processes.

Allocate at least 15% of any AI initiative budget to change management, workforce development, and stakeholder engagement rather than treating these as items to cut when budgets tighten. This investment pays for itself through higher adoption rates, fewer costly corrections after deployment, and sustainable outcomes rather than pilot projects that never scale. The agencies achieving value from AI are those treating human transformation as equally important to technical implementation, recognizing that AI capabilities that people don't use or don't use effectively deliver nothing regardless of their algorithmic sophistication.

AI has finally given government leaders a realistic alternative to decade-long rip-and-replace projects that consume enormous resources, while delivering uncertain outcomes. Composable extensions, low-code orchestration, and ERP-native AI capabilities enable agencies to modernize in measured sprints rather than massive transformations. Yet the same forces that flatten timelines also multiply hidden risks of bias, lock-in, and privacy violations that can erode public confidence overnight. Independent, vendor-neutral governance is therefore not a luxury but the foundation on which AI's public value depends. The agencies that pair human-centered oversight with AI's computational power will spend smarter, serve constituents faster, and rebuild the trust that large-scale technology failures have too often undermined. Those that sprint alone risk repeating Birmingham's costly lesson under far less forgiving public scrutiny.

12

The Complete Roadmap:
Your Step-by-Step Guide
to Breaking Free

WHAT YOU'LL LEARN

This chapter synthesizes everything we've explored into a practical, step-by-step framework that moves from abstract principle to concrete action:

- You'll find a five-phase roadmap that takes you from initial leadership alignment through sustained value realization, with clear objectives, activities, and deliverables for each phase.
- You'll understand how to adapt this methodology to your organization's unique culture, complexity, and constraints, while maintaining the disciplined approach that distinguishes successful modernizations from expensive failures.

- You'll leave with a specific action plan for the next 48 hours, because the window for establishing independence is measured in months rather than years, and every day of delay strengthens vendor leverage, while weakening your position.

THE URGENCY BEHIND THE ROADMAP

We began this journey with Birmingham's £216 million wake-up call, a story of good intentions and even proven technology undone by a process that surrendered control to parties whose interests diverged from the council's own. Across these chapters, we've explored the compounding risks of legacy systems, the critical importance of Phase Zero preparation, the transformative potential of AI when properly governed, and the nonnegotiable requirement for genuine independence at every stage. We've seen how easily projects drift into disaster when they lack clear vision and empowered structures to protect that vision against the pressures that inevitably arise.

The contrast between failure and success isn't theoretical. While Birmingham consumed hundreds of millions with vendor-led chaos, Monroe County completed their ERP modernization within budget and ahead of schedule, achieving adoption rates that validated every investment in preparation and change management. While other agencies struggled through multiyear implementations that delivered systems nobody wanted to use, the organizations that maintained independence through disciplined advisory relationships achieved outcomes that justified the political capital their leaders expended.

What's changed since these stories began unfolding is the acceleration of every pressure we've discussed. The legacy systems supporting your operations are one day older, one day more fragile, one day closer to the failure that will force crisis-mode decision-making under the worst possible conditions. Your most experienced staff, the ones who understand those aging platforms intimately, are one day closer to retirement, and the institutional knowledge they carry cannot be transferred through documentation or training because much of it exists only in their heads. The cyber threats targeting government systems have evolved from theoretical risks to weekly occurrences, with ransomware attacks crippling operations for weeks at a time. And the gap between citizen expectations shaped by consumer technology and the experience your current systems deliver grows wider with every Amazon order tracked in real-time, while business license renewals require faxed documents and weeks of waiting.

The vendors understand these pressures better than anyone, and they've positioned themselves to benefit from your urgency. They've built business models around complexity that requires their ongoing involvement, certification programs that create artificial scarcity of qualified implementers, upgrade paths that are mandatory and expensive, and contract structures that make data extraction difficult and costly. This isn't conspiracy but a rational business strategy executed with discipline over decades. They've created ecosystems where dependency deepens over time rather than diminishing.

However, their leverage depends entirely on your belief that no alternatives exist—the moment you recognize that you can take control, build internal capabilities, and utilize independent expertise that serves your interests rather than vendor interests, their negotiating position weakens dramatically. The roadmap that follows is the practical framework for making that recognition operational.

THE FIVE PHASES OF BREAKING FREE

This methodology is not a rigid prescription but a proven framework that successful organizations adapt to their unique circumstances. The phases build on each other sequentially, with each phase's deliverables creating the foundation for the next. Skipping phases or rushing through them to save time inevitably costs more time later when the gaps they would have addressed create problems that must be solved under pressure.

Monroe County followed this path. So did the other success stories we've examined. The failures we've studied share a common characteristic: They treated early phases as optional or delegated them to vendors whose interests favored speed over thoroughness.

Phase 1: Foundation

The objective of this phase that lasts 2 to 3 months is to secure genuine leadership alignment and establish the governance structures that will guide the entire initiative. Before you can productively think about software, you must build the human and political architecture for success.

Begin by forming the Executive Steering Committee and formally appointing the Executive Sponsor. As we explored in Chapter 8, the Sponsor is the single most important role in the entire initiative, the ultimate decision-maker with authority to break ties and the political capital to enforce difficult choices for the good of the whole organization. This cannot be a ceremonial appointment or a delegation to someone without real power. The Sponsor must have direct access to the chief executive, budget authority sufficient to the initiative's scope, and willingness to invest significant personal time in oversight rather than treating the role as one more committee membership requiring occasional attendance.

With leadership structure established, the next critical action is selecting your independent advisory partner. This decision warrants the same rigor you would apply to selecting the eventual software vendor, because the advisor will shape every subsequent decision and their independence, or lack thereof, will determine whether you receive counsel that serves your interests or counsel that serves other agendas. Focus on firms with deep public sector experience and verifiable track records of independence, examining not just their marketing claims but their actual client relationships, revenue sources, and history of recommending against implementations when circumstances warranted.

Once your advisor is engaged, work together to draft and ratify the Governance Charter. This document serves as the initiative's constitution, signed by the Executive Sponsor and every Steering Committee member, formalizing decision rights, escalation paths, and accountability structures. It prevents the gridlock that dooms initiatives when nobody has clear authority to resolve disputes. It also creates the documentation that protects everyone involved when auditors or oversight bodies later examine how decisions were made.

The phase concludes with initial risk assessment that identifies major political, financial, and operational threats before they have opportunity to materialize. This assessment should be brutally honest about organizational weaknesses, past project failures, stakeholder resistance likely to emerge, and external factors that could disrupt progress. The risks you identify now can be mitigated through planning; the risks you discover mid-implementation become crises that consume resources and derail timelines.

The critical deliverable from this phase is the signed Governance Charter with accompanying risk register. Without these artifacts, you lack the foundation for everything that follows.

Phase 2: Discovery and Vision

This phase lasts 4 to 8 months and represents the heart of Phase Zero, the preparation work that most organizations skip to their eventual regret. The objective is completing the essential deliverables detailed in Chapter 4, creating a business case grounded in evidence and a shared vision for the future state before engaging software vendors whose presence inevitably shapes and constrains the conversation.

Guided by your independent partner, your team embarks on discovery that employs modern tools to understand your actual current state. Process-mining software can analyze system logs to reveal how work actually flows, identifying bottlenecks, workarounds, and variations that formal documentation doesn't capture. This evidence-based understanding prevents the common failure mode of building new systems around assumptions about current operations that prove incorrect.

Simultaneously, run stakeholder engagement and visioning workshops that accomplish multiple objectives beyond simple requirements gathering. These sessions build consensus around the future state, surface concerns that will become resistance if left unaddressed, and create shared ownership that sustains commitment through the inevitable challenges ahead. The workshops should include not just department heads and power users but representatives from across the organization who will live with whatever decisions emerge.

To prevent your vision from being constrained by past experience, your independent advisor should arrange "Art of the Possible" educational sessions where vendors demonstrate capabilities without the pressure of competitive evaluation. These non procurement demonstrations open eyes to what modern technology can accomplish, breaking teams out of the mindset that unconsciously assumes the future must resemble the past. The goal is informed ambition rather than either unrealistic expectations or unnecessarily modest aspirations.

The culmination of this work is the Phase Zero Findings Report, a comprehensive document packaging all essential deliverables into a single source of truth. This report provides the data-backed foundation for procurement, the baseline against which implementation progress will be measured, and the artifact that documents the preparation work for stakeholders and oversight bodies who will later want to understand how decisions were made.

Phase 3: Procurement and Selection

With solid foundation in place, you're ready to engage the market from a position of strength rather than dependency. This phase lasts 4 to 6 months during which a disciplined selection process is run, designed to identify not just capable software but the right long-term partner for your organization's specific needs and culture.

Begin with RFP development that leverages the rich outputs of Phase Zero to ask insightful, scenario-based questions rather than generic feature checklists that every vendor can claim to satisfy. The requirements you've developed, grounded in your actual operational needs and validated through stakeholder engagement, become the framework for evaluation rather than vendor-supplied templates that inevitably favor the vendor's strengths.

The most revealing element of selection is scenario-based demonstration, what practitioners sometimes call "bake-offs," where vendors must prove their systems handle your most complex processes using your own sanitized data rather than rehearsed presentations optimized for visual effect. These sessions reveal how systems actually work rather than how vendors describe them, and they expose integration challenges, usability issues, and capability gaps that polished demonstrations conceal.

While vendors prepare demonstrations, your independent advisor conducts reference checking that goes beyond the curated lists vendors provide. Conversations with organizations the vendor didn't suggest, particularly those with similar scale, complexity, or requirements, reveal the unvarnished truth about real-world performance, support responsiveness, and implementation challenges.

Contract negotiation should benchmark proposed terms against industry standards and structure payments around performance milestones rather than arbitrary timelines. Your advisor's experience across multiple engagements provides the comparative knowledge necessary to identify provisions that seem standard but actually favor vendors, contingency language that provides inadequate protection, and support terms that leave you exposed when problems arise.

The phase concludes with a signed performance-based contract that aligns vendor success directly with your outcomes. This document operationalizes the independence principles that have guided the process, creating contractual protection for the position you've worked to establish.

Phase 4: Implementation and Adoption

This is the longest and most intensive phase, lasting 18 to 36 months. This the stage where vision becomes operational reality. The objective is managing technical implementation, while simultaneously driving user adoption through deeply integrated organizational change management, because systems that work technically but that people don't use or don't use effectively deliver nothing.

Your independent advisor continues operating the Project Management Office, providing verification and validation of the system integrator's work that the integrator cannot credibly provide about themselves. This oversight catches problems early when they're addressable, and it maintains the accountability that prevents the drift toward vendor convenience characterizing unsupervised implementations.

The organizational change management strategy developed in Phase Zero now executes in full. Change champions throughout the organization share accurate information and address concerns before they become resistance. Strategic communications maintain awareness and manage expectations. Readiness assessments before major milestones identify gaps requiring intervention before they derail progress.

Deploy AI-enabled training tools, the Digital Adoption Platforms discussed in Chapter 9, that provide guidance within applications at the moment of need rather than relying on classroom sessions that people forget before they encounter real work. This approach compresses time to proficiency and reduces the support burden that overwhelms help desks during transition periods.

Governance dashboards with predictive monitoring provide leadership visibility into risks before they become crises. The patterns that precede problems, whether schedule slippage, budget pressure, or adoption shortfalls, become visible early enough for intervention rather than late when damage has occurred.

The phase concludes not merely with technical go-live but with a stabilization period during which business value and user adoption are formally measured against the targets established in Phase Zero. The contract provisions negotiated in Phase 3 should tie final payments to achieving these targets, maintaining vendor accountability through the period when their attention naturally shifts toward the next sale.

Phase 5: Value Realization and Continuous Improvement

Breaking free is not a destination reached at go-live but a sustained commitment to operating differently than you did before. This ongoing phase ensures the initiative delivers the long-term value promised in the business case and builds the organizational capability for continuous improvement that genuine independence requires.

Active benefits tracking uses the metrics established in Phase Zero to demonstrate return on investment to stakeholders who approved the initiative and to build the evidence base supporting future investments. This documentation matters politically because it answers the inevitable questions about whether the investment was worthwhile, and it matters practically because it identifies where expected benefits haven't materialized and intervention may be required.

Transition project governance structures into a permanent Center of Excellence, an internal team responsible for ongoing system management and optimization. This transition represents the culmination of capability building that should have occurred throughout implementation, developing staff expertise that enables self-sufficiency rather than perpetual vendor dependency for every enhancement or issue resolution.

With stable foundation in place, the composable architecture and rapid development tools discussed in Chapter 11 enable quick-iteration solutions to problems that once required expensive, lengthy vendor engagements. Departments can address specific operational needs in weeks using low-code platforms and AI-enabled tools, building organizational capability with each solution deployed rather than deepening vendor relationships.

The capstone is periodic value reporting that documents initiative success against original commitments, closing the loop of accountability and demonstrating responsible stewardship. These reports serve both internal governance and external transparency, providing the evidence that supports future investments and the documentation that satisfies oversight requirements.

YOUR 48-HOUR ACTION PLAN

The described roadmap provides direction, but direction without action is merely theory. The vendors strengthening their position while you consider your options aren't waiting, and neither should you. Within the next 48 hours, take these specific actions:

1. Inventory your three highest-risk systems by asking which platforms would cause the most severe operational disruption if they failed tomorrow. For each, document who maintains them, what happens when that person is unavailable, what the recovery plan is if the system becomes corrupted or inaccessible, and when the last comprehensive backup was tested through actual restoration. This inventory creates the urgency case for leadership attention.

2. Audit your current vendor dependencies by examining, for each major system, who controls the source code, where your data physically resides, what the contractual provisions are for data extraction if you change vendors, and what the cost has been for enhancements and support over the past 3 years. This audit reveals the leverage vendors currently hold and the cost of continuing on the current path.

3. Schedule an independent assessment by identifying qualified advisory firms without vendor partnerships and requesting initial conversations about your situation. These conversations cost nothing and provide perspective on your options that you cannot obtain from parties with interests in particular outcomes. The assessment should evaluate not just technology but governance readiness,

stakeholder alignment, and organizational capability for the initiative ahead.

4. Assemble the leadership conversation by identifying who in your organization has authority to initiate a modernization initiative and requesting time to present what you've learned. Bring this roadmap, your risk inventory, your dependency audit, and the questions that emerged from your independent conversations. The goal is not immediate decision but informed discussion about whether current trajectory is acceptable and what alternatives deserve exploration.

The revolution in government technology that AI enables and that independence makes possible starts with one decision—to stop accepting that vendor control is inevitable and to take the first step toward a different outcome. The organizations that made this decision and followed through are the success stories we've examined throughout this book. The organizations that recognized the problems but never acted are the cautionary tales.

Which story will your organization tell?

The path is clear. The methodology is proven. The only remaining question is whether you'll take the first step.

Appendix

Templates and Checklists

The templates in this appendix are the practical tools that translate this book's principles into action. They're organized into two tiers: foundational templates included here in full, and advanced tools available through our online resource center.

Why the distinction?

The foundational templates give you everything you need to start your journey with discipline and structure. The advanced tools—weighted evaluation scorecards, proprietary assessment frameworks, and detailed negotiation guides—reflect methodology refined through hundreds of government engagements. These are available to readers who want to go deeper, and of course, we bring the full toolkit when engaged as your advisory partner.

Download editable versions of all templates at **STARTATZERO. AVEROADVISORS.COM.**

FOUNDATIONAL TEMPLATES (Included in Full)

1. PHASE ZERO 100-DAY CHECKLIST

This checklist guides your team through the essential Phase Zero activities discussed in Chapter 4. Customize timelines and owners based on your organization's complexity.

Day/Week	Activity	Owner	Status	Notes
Days 1–7	Appoint Phase Zero Lead and form core team (3–5 members representing IT, finance, HR, operations).	Executive Sponsor	☐ To Do ☐ In Progress ☐ Done	Identify internal experts or independent consultant.
Days 8–14	Conduct kickoff meeting: align on goals, review business case, assign roles.	Phase Zero Lead	☐ To Do ☐ In Progress ☐ Done	Consider AI-assisted process scanning tools.
Weeks 3–4	Document current state processes: Map 10--5 key workflows using observation, not procedure manuals.	Core Team	☐ To Do ☐ In Progress ☐ Done	Capture workarounds and exceptions.

Day/Week	Activity	Owner	Status	Notes
Weeks 5–6	Create stakeholder power map: Identify official and unofficial influencers; conduct 20+ interviews.	Phase Zero Lead	☐ To Do ☐ In Progress ☐ Done	Document potential resistance.
Weeks 7–8	Complete integration inventory and data quality assessment: List all system connections; assess data cleanliness.	IT Lead	☐ To Do ☐ In Progress ☐ Done	Flag duplicates, errors, gaps.
Weeks 9–10	Build compliance requirements matrix and risk register: Document regulations, union rules, audit requirements.	Core Team	☐ To Do ☐ In Progress ☐ Done	Include AI governance requirements.
Weeks 11–12	Conduct change readiness assessment and organizational design review: Survey readiness; identify future roles.	HR Lead	☐ To Do ☐ In Progress ☐ Done	Identify skill gaps; plan training.

Day/ Week	Activity	Owner	Status	Notes
Weeks 13–14	Develop business case and benefits plan: Define measurable benefits with owners and timelines.	Finance Lead	☐ To Do ☐ In Progress ☐ Done	Establish baseline metrics for ROI.
Day 100	Go/No-Go decision meeting: Present findings; decide whether to proceed to procurement.	Executive Sponsor	☐ To Do ☐ In Progress ☐ Done	Compile Phase Zero Findings Report

2. INDEPENDENT PMO CHARTER

This charter establishes your Independent Project Management Office. Customize sections in brackets.

INDEPENDENT PMO CHARTER

Project Name: [ERP Modernization Project] **Date:** [Date]

Purpose
To provide unbiased oversight ensuring the project delivers on time, on budget, and aligned with agency interests, free from vendor influence.

Scope

The Independent PMO will oversee all phases from Phase Zero through post-go-live stabilization, verify deliverables against Phase Zero requirements, manage risks and change orders, facilitate escalations, and provide objective status reporting.

Authority

The PMO has direct access to all project documents, meetings, and systems. The PMO has the right to halt activities presenting unmitigated risk. The PMO reports directly to the Executive Sponsor, not to IT or the vendor.

Roles and Responsibilities

Role	Name	Responsibilities
PMO Lead	[Name]	Daily oversight, status reporting, risk management
Independent Advisor	[Firm/ Name]	Unbiased reviews, recommendations, IV&V

Independence Attestation

All PMO members attest they have no financial relationships with vendors or system integrators under consideration or contract.

Reporting Cadence

Weekly dashboards to project team; bi-weekly briefings to Steering Committee; monthly executive summaries to sponsor.

Success Metrics

Metric	Target
On-time deliverables	95%
Budget variance	<10%
User adoption at go-live +90 days	90%

Signatures

Role	Signature	Date
Executive Sponsor	______________________	
PMO Lead	______________________	

3. THREE-LAYER RACI MATRIX

This matrix clarifies responsibilities across the vendor, system integrator, and independent PMO (see the three-layer model discussed in Chapter 10).

R = Responsible (does the work) |
A = Accountable (approves) |
C = Consulted (input) |
I = Informed (updated)

Task/ Activity	Software Vendor	System Integrator	Independent PMO	Executive Sponsor
Define requirements	C	C	R (validates)	I
Develop RFP	I	I	R (drafts)	A (approves)
Evaluate vendors	—	—	R (facilitates)	A (selects)
Submit change order	R (requestor)	R (requestor)	C (reviews)	I
Approve change order	I	I	R (validates cost/need)	A (final sign-off)
Conduct UAT	C (fixes defects)	R (executes)	A (certifies quality)	I
Go-live decision	C (readiness input)	C (readiness input)	R (facilitates)	A (the decider)
Post-go-live support	R (provides)	R (provides)	C (monitors)	I

4. SCOPE CHANGE REQUEST FORM

Use this form to evaluate and approve scope changes, preventing the uncontrolled scope creep that doomed Birmingham and countless other projects.

Scope Change Request

Request ID: [SCR-001] **Date Submitted:** [Date]
Submitted By: [Name/Role]

Description of Proposed Change

[Detailed description of what is being requested and why it is needed]

Business Justification

[Why this change is necessary; what problem it solves; consequence of not implementing]

Impact Assessment

Impact Area	Assessment
Schedule	[Days added/removed; affected milestones]
Budget	[Cost increase/decrease; source of funds]
Risk	[New risks introduced; mitigation required]
Benefits	[Impact on business case benefits]
Resources	[Additional staff/effort required]

Alternatives Considered

[Other options evaluated, including deferral to future phase]

Independent PMO Assessment

Assessment Area	PMO Evaluation
Is this change necessary?	
Is the cost estimate reasonable?	
Is the timeline impact accurate?	
Recommendation	☐ Approve ☐ Approve with modifications ☐ Defer ☐ Reject
Rationale	

Approvals

Role	Decision	Signature	Date
Project Manager	☐ Approve ☐ Reject	__________	
Independent PMO	☐ Approve ☐ Reject	__________	
Executive Sponsor (if >$____)	☐ Approve ☐ Reject	__________	

Final Status: ☐ Approved ☐ Rejected ☐ Deferred to: __________

5. PARALLEL RUN CUTOVER CHECKLIST

Use this checklist when running legacy and new systems in parallel during the cutover period.

Step	Activity	Owner	Status	Notes
1	Define parallel run duration and success criteria.	PMO Lead	☐ To Do ☐ In Progress ☐ Done	Typical: 1–2 pay periods or month-end cycles
2	Set up synchronized environments (legacy and new).	IT Team	☐ To Do ☐ In Progress ☐ Done	Ensure data flows to both systems.
3	Train super users on parallel monitoring procedures.	Training Lead	☐ To Do ☐ In Progress ☐ Done	Focus on reconciliation techniques.
4	Execute daily data reconciliation.	IT/Vendor	☐ To Do ☐ In Progress ☐ Done	Flag discrepancies >1%
5	Monitor key transaction accuracy metrics.	Independent PMO	☐ To Do ☐ In Progress ☐ Done	Target: 99%+ match rate
6	Resolve discrepancies in new system only.	SI Team	☐ To Do ☐ In Progress ☐ Done	Document all fixes.

Step	Activity	Owner	Status	Notes
7	Conduct stakeholder readiness confirmation.	Department Heads	☐ To Do ☐ In Progress ☐ Done	Formal sign-off required
8	Execute cutover to new system.	IT Team	☐ To Do ☐ In Progress ☐ Done	Rollback plan ready
9	Conduct post-cutover audit (1 week).	Independent PMO	☐ To Do ☐ In Progress ☐ Done	Confirm 95%+ adoption.
10	Decommission legacy system access.	IT Team	☐ To Do ☐ In Progress ☐ Done	Archive per retention policy

6. VENDOR TERMINATION LETTER TEMPLATE

Use this template when vendor performance requires contract termination. Consult legal counsel before sending.

[Agency Letterhead]
[Date]
[Vendor Name] [Vendor Address]

Subject: Termination of Contract for [Project Name/Service]
Contract Number: [Number]

Dear [Vendor Contact Name]:

This letter provides formal notice of termination of the contract between [Agency Name] and [Vendor Name] for [project/service description], effective [termination date, typically 30 days from receipt].

This decision is based on the following material breaches of contract:

[Specific breach with contract reference; e.g., "Failure to meet Milestone 3 deliverables per Section 5.2, now 60+ days overdue"]

[Second breach; e.g., "Budget overrun exceeding 20% without required approval per Section 7.1"]

[Third breach, e.g.; "Failure to provide qualified personnel as specified in Attachment B"]

Per Section [X] of the contract, we invoke our right to terminate for cause. You are required to cease all work immediately upon the effective date

Deliver all work products, documentation, and agency data within [X] business days

Provide transition assistance as specified in Section [Y]

Refund unearned fees in the amount of $[Amount]

Failure to comply with these requirements may result in legal action to recover damages. Please direct all transition-related questions to [Contact Name] at [phone/email].

Sincerely,

[Name] [Title] [Agency]

cc: [Legal Counsel, Procurement Officer, Executive Sponsor]

FRAMEWORK DESCRIPTIONS

The following tools are essential to successful modernization but require customization to your specific context. Descriptions are included here so you understand what good looks like. Full templates are available at **STARTATZERO.AVEROADVISORS.COM**, and we bring enhanced versions with benchmarking data when engaged as your advisory partner.

7. EXECUTIVE STEERING COMMITTEE GOVERNANCE CHARTER

The Governance Charter is arguably the most important document you'll create in Phase 1. It transforms verbal commitments into binding agreements and prevents the authority vacuum that allows projects to drift.

What It Should Contain

- *Vision and Objectives*: clear statement of what success looks like and the measurable outcomes the project must achieve. These become the criteria against which all decisions are evaluated.

- *Governance Structure*: Named individuals with specific roles—Executive Sponsor (the ultimate decision-maker), Steering Committee members (representing key functional areas), and the Independent Advisor relationship. Each role needs defined time commitments, not just titles.

- *Decision Rights Matrix*: The most critical section. It includes information such as who can approve what and at what thresholds do decisions escalate. This prevents the paralysis that occurs when nobody knows who has authority and the chaos that occurs when everyone thinks they do. Typical categories include scope changes, budget adjustments, timeline modifications, vendor disputes, and risk acceptance.

- *Escalation Procedures:* These include a clear path from project team to PMO to Steering Committee to Executive Sponsor, with defined timeframes for resolution at each level. Issues that languish unresolved kill projects.

- *Meeting Cadence*: Specifies frequency for Steering Committee meetings, sponsor briefings, and stakeholder communications. Projects that "meet when needed" rarely meet when they should.

- *Success Metrics*: Describe how you'll measure whether the project is on track—schedule adherence, budget variance, scope stability, adoption rates, and benefits realization.

- *Signatures*: Every Steering Committee member and the Executive Sponsor sign, thus creating accountability that verbal agreements lack.

8. GO/NO-GO DECISION CRITERIA

At critical junctures—end of Phase Zero, pre-go-live, end of stabilization—you need objective criteria to decide whether to proceed. Subjective judgment ("it feels ready") leads to disasters; structured assessment protects everyone.

STRUCTURE YOUR CRITERIA AROUND

1. Technical Readiness

- Are critical defects resolved?
- Has performance testing confirmed the system meets requirements?
- Are integrations working?
- Has data migration been validated?
- Is the rollback plan tested?

2. User Readiness

- Is training complete?
- Have users demonstrated proficiency?
- Are super users certified and available?
- Is the help desk staffed for the surge?

3. Organizational Readiness

- Has leadership formally confirmed readiness?
- Are department heads prepared?
- Is the Change Champion network activated?
- Have go-live communications been sent?

4. Operational Readiness

- Is the cutover plan finalized?
- Was the parallel run successful?
- Is vendor support confirmed for go-live weekend?

Decision Framework

Each criterion should be rated GO (meets threshold), CONDITIONAL GO (meets threshold with documented mitigation), or NO-GO (does not meet threshold). Designate certain criteria as *critical*—a single NO-GO on a critical item halts the decision until resolved.

This framework removes politics from the decision. When the Executive Sponsor asks "are we ready?" the answer is documented evidence, not opinion.

9. STAKEHOLDER POWER/INFLUENCE MAP

Chapter 9 emphasized that stakeholder alignment determines whether systems get used or rejected. A stakeholder map makes the invisible visible, identifying where you need to invest relationship capital.

The Framework

Map each stakeholder on two dimensions:

1. Their *power* to influence the project's success or failure
2. Their *interest* level in the project's outcome.

This creates four quadrants requiring different strategies (Figure A.1):

STAKEHOLDER ENGAGEMENT MATRIX

HIGH INTEREST

INTEREST

Keep Satisfied
(High Power/Low Interest)

Manage Closely
(High Power/High Interest)

POWER

HIGH POWER

LOW POWER

Monitor
(Low Power/Low Interest)

Keep Informed
(Low Power/High Interest)

POWER

INTEREST

LOW INTEREST

Figure A.1 Stakeholder Engagement Matrix

- *High Power / High Interest*: Manage them closely. These stakeholders can make or break your project and care deeply about the outcome. Regular engagement, involvement in decisions, and proactive communication are essential.

- *High Power / Low Interest*: Keep them satisfied. They have authority but aren't paying close attention—until something goes wrong. Keep them informed without overwhelming them; ensure no surprises reach them.

- *Low Power / High Interest*: Keep them informed. They care deeply but have limited formal authority. Often includes end users who can create grassroots resistance or support. Regular communication and feedback channels matter.

- *Low Power / Low Interest*: Monitor them. Minimal engagement required, but watch for changes that might move them to another quadrant.

Beyond the Grid

For high-power stakeholders, document

- Their current position (champion to opponent)
- What they care about most
- What they fear
- Who influences them
- Your strategy for moving them toward support
- Don't forget informal influencers, the long-tenured employees without titles who shape how their colleagues perceive change.

10. CHANGE CHAMPION NETWORK

Change Champions are your grassroots adoption force—respected employees who bridge the gap between the project team and end users. Chapter 9 explained why vendor-delivered training fails. Champions are part of the solution.

Network Design

Target one Champion per 25–50 end users, representing every major functional area and location. Selection criteria matter more than titles:

Look for informal influence, positive attitude, good communication skills, and willingness to dedicate 2–4 hours weekly.

Champion Responsibilities

- *Communication*: Cascade project information to peers, dispel rumors, explain the *why* behind changes
- *Feedback*: Gather concerns from colleagues, report resistance patterns, serve as early warning system
- *Support*: Complete training early, provide peer coaching at go-live, help struggling colleagues
- *Validation*: Participate in User Acceptance Testing, confirm the system works for real-world needs

What Champions Are Not

Project team members: They keep their day jobs.

- *Help desk*: They escalate technical issues.
- *Decision-makers*: They influence and advise.
- *Expected to have all answers*: They know where to find them.

Formalize the network with a charter that defines expectations, provides support (early training access, recognition), and secures manager approval for the time commitment.

ADVANCED TOOLS (Available Online)

The following tools are available to readers at **STARTATZERO. AVEROADVISORS.COM**. Registration is required.

11. VENDOR EVALUATION SCORECARD

It is weighted scoring framework for objectively comparing vendor finalists across functional fit, technical architecture, implementation approach, vendor viability, and total cost of ownership. Includes guidance on weight allocation based on organizational priorities and scoring calibration across evaluation team members.

12. INDEPENDENT ADVISOR SELECTION CRITERIA

Before you select your ERP vendor, you need to select the advisor who will guide that process. This evaluation framework covers independence verification (pass/fail criteria that disqualify conflicted firms), public sector experience assessment, service capability evaluation, and team qualification review.

13. VENDOR REFERENCE CHECK INTERVIEW GUIDE

A structured interview protocol for conducting reference checks that go beyond vendor-provided contacts. Includes questions designed to surface implementation timeline accuracy, budget variance history, change order patterns, post-go-live support quality, and the ultimate question: "Knowing what you know now, would you select this vendor again?"

14. CONTRACT NEGOTIATION CHECKLIST

A comprehensive checklist covering payment structure (milestone-based with holdback), data ownership and portability, intellectual property rights, service level agreements, termination provisions, staffing commitments, pricing protection, and security requirements. Includes AI-specific provisions for contracts involving embedded AI capabilities.

15. BENEFITS-REALIZATION TRACKER

A framework for tracking whether your project delivers the value promised in the business case, from financial benefits (cost savings, efficiency gains) through operational improvements (cycle time reduction, error reduction) to strategic outcomes (real-time visibility, self-service capability). Includes ROI calculation methodology and variance analysis structure.

16. AI GOVERNANCE AND READINESS ASSESSMENT

Based on the Avèro AI Maturity Model, this assessment evaluates organizational readiness across nine dimensions spanning People (skills, leadership, dedicated team), Process (governance, transparency, compliance), and Technology (security, infrastructure, data quality). Identifies your current maturity level and gaps requiring attention before AI deployment.

Getting the Full Toolkit

Visit **STARTATZERO.AVEROADVISORS.COM** to access editable versions of all templates in this appendix (Word and Excel formats)

- The advanced tools described above
- Video walkthroughs explaining how to use each template effectively
- Updates as methodologies evolve and new tools become available

For organizations seeking hands-on support applying these frameworks to your specific situation, Avèro Advisors brings enhanced versions of these tools along with benchmarking data from hundreds of government engagements and the expertise to customize them to your unique context.

www.ingramcontent.com/pod-product-compliance
Lightning Source LLC
Chambersburg PA
CBHW071733150726
47998CB00005B/1623